The Handbook of Environmental Chemistry

Volume 2   Part D

Edited by O. Hutzinger

Reactions and Processes

With Contributions by
P. B. Barraclough, N. O. Crossland, R. Herrmann,
W. Mabey, C. M. Menzie, Th. Mill,
P. B. Tinker, M. Waldichuk, C. J. M. Wolff

With 47 Figures

Springer-Verlag
Berlin Heidelberg GmbH

Professor Dr. Otto Hutzinger

University of Bayreuth
Chair of Ecological Chemistry and Geochemistry
Postfach 10 12 51, D-8580 Bayreuth
Federal Republic of Germany

ISBN 978-3-662-15122-8

Library of Congress Cataloging in Publication Data
The Handbook of environmental chemistry.
Includes bibliographies and indexes.
Contents: v. 1. The natural environment and the biogeochemical cycles/with contributions by P. Craig...
[et al.] – v. 2. Reactions and processes/with contributions by W. A. Bruggeman... [et al.] – v. 3. Anthropogenic
compounds/with contributions by R. Anliker... [etc.]
1. Environmental chemistry. I. Hutzinger, O.
QD31.H335   574.5′222   80-16607
ISBN 978-3-662-15122-8   ISBN 978-3-540-39460-0 (eBook)
DOI 10.1007/978-3-540-39460-0

© Springer-Verlag Berlin Heidelberg 1988
Originally published by Springer-Verlag Berlin Heidelberg New York in 1988
Softcover reprint of the hardcover 1st edition 1988

2152/3140-543210

Preface

Environmental Chemistry is a relatively young science. Interest in this subject, however, is growing very rapidly and, although no agreement has been reached as yet about the exact content and limits of this interdisciplinary subject, there appears to be increasing interest in seeing environmental topics which are based on chemistry embodied in this subject. One of the first objectives of Environmental Chemistry must be the study of the environment and of natural chemical processes which occur in the environment. A major purpose of this series on Environmental Chemistry, therefore, is to present a reasonably uniform view of various aspects of the chemistry of the environment and chemical reactions occurring in the environment.

The industrial activities of man have given a new dimension to Environmental Chemistry. We have now synthesized and described over five million chemical compounds and chemical industry produces about one hundred and fifty million tons of synthetic chemicals annually. We ship billions of tons of oil per year and through mining operations and other geophysical modifications, large quantities of inorganic and organic materials are released from their natural deposits. Cities and metropolitan areas of up to 15 million inhabitants produce large quantities of waste in relatively small and confined areas. Much of the chemical products and waste products of modern society are released into the environment either during production, storage, transport, use or ultimate disposal. These released materials participate in natural cycles and reactions and frequently lead to interference and disturbance of natural systems.

Environmental Chemistry is concerned with *reactions in the environment*. It is about distribution and equilibria between environmental compartments. It is about reactions, pathways, thermodynamics and kinetics. An important purpose of this Handbook is to aid understanding of the basic distribution and chemical reaction processes which occur in the environment.

Laws regulating toxic substances in various countries are designed to assess and control risk of chemicals to man and his environment. Science can contribute in two areas to this assessment; firstly in the area of toxicology and secondly in the area of chemical exposure. The available concentration ("environmental exposure concentration") depends on the fate of chemical compounds in the environment and thus their distribution and reaction behaviour in the environment. One very important contribution of Environmental Chemistry to the above mentioned toxic substances laws is to develop laboratory test methods, or mathematical correlations and models that predict the environmental fate of new chemical compounds. The third purpose of this Handbook is to help in the basic understanding and development of such test methods and models.

The last explicit purpose of the handbook is to present, in a concise form, the most important properties relating to environmental chemistry and hazard assessment for the most important series of chemical compounds.

At the moment three volumes of the Handbook are planned. Volume 1 deals with the natural environment and the biogeochemical cycles therein, including some background information such as energetics and ecology. Volume 2 is concerned with reactions and processes in the environment and deals with physical factors such as transport and adsorption, and chemical, photochemical and biochemical reactions in the environment, as well as some aspects of pharmacokinetics and metabolism within organisms. Volume 3 deals with anthropogenic compounds, their chemical backgrounds, production methods and information about their use, their environmental behaviour, analytical methodology and some important aspects of their toxic effects. The material for volumes 1, 2, and 3 was more than could easily be fitted into a single volume, and for this reason, as well as for the purpose of rapid publication of available manuscripts, all three volumes are published as a volume series (e.g. Vol. 1; A, B, C). Publisher and editor hope to keep the material of the volumes 1 to 3 up to date and to extend coverage in the subject areas by publishing further parts in the future. Readers are encouraged to offer suggestions and advice as to future editions of "The Handbook of Experimental Chemistry".

Most chapters in the Handbook are written to a fairly advanced level and should be of interest to the graduate student and practising scientist. I also hope that the subject matter treated will be of interest to people outside chemistry and to scientists in industry as well as government and regulatory bodies. It would be very satisfying for me to see the books used as a basis for developing graduate courses on Environmental Chemistry.

Due to the breadth of the subject matter, it was not easy to edit this Handbook. Specialists had to be found in quite different areas of science who were willing to contribute a chapter within the prescribed schedule. It is with great satisfaction that I thank all authors for their understanding and for devoting their time to this effort. Special thanks are due to the Springer publishing house and finally I would like to thank my family, students and colleagues for being so patient with me during several critical phases of preparation for the Handbook, and also to some colleagues and the secretaries for their technical help.

I consider it a privilege to see my chosen subject grow. My interest in Environmental Chemistry dates back to my early college days in Vienna. I received significant impulses during my postdoctoral period at the University of California and my interest slowly developed during my time with the National Research Council of Canada, before I was able to devote my full time to Environmental Chemistry in Amsterdam. I hope this Handbook will help deepen the interest of other scientists in this subject.

Otto Hutzinger

This preface was written in 1980. Since then publisher and editor have agreed to expand the Handbook by three new open-ended volume series: Air Pollution, Water Pollution and Environmental Modelling. These broad topics could not be fitted easily into the headings of the first three volumes.

All six volume series will be integrated through the choice of topics covered and by a system of cross referencing.

The outline of the Handbook is thus as follows:
1. The Natural Environment and the Biogeochemical Cycles
2. Reactions and Processes
3. Anthropogenic Compounds
4. Air Pollution
5. Water Pollution
6. Environmental Modelling

Bayreuth, March 1988 Otto Hutzinger

Contents

List of Contributors

Dr. Peter B. Barraclough
Rothamsted Experimental Station
Harpenden, Herts AL5 2JQ
United Kingdom

Dr. N. O. Crossland
Shell Research Limited
Sittingbourne Research Centre
Sittingbourne, Kent ME9 8AG
United Kingdom

Prof. Dr. Reimer Herrmann
Lehrstuhl für Hydrologie
Universität Bayreuth
Postfach 101251
D-8580 Bayreuth
Federal Republic of Germany

Dr. William Mabey
Kennedy/Jenks Engineers
San Francisco, CA 94103, USA

Dr. Calvin M. Menzie
American University
Washington, D.C., USA

Dr. Theodore Mill
Chemistry Laboratory
SRI International
333 Ravenswood Avenue
Menlo Park, CA 94025, USA

Dr. Philip Bernard Tinker
Natural Environment Research Council
Terrestrial and Freshwater Sciences
Polaris House, North Star Avenue
Swindon Wilts. SN2 1EU
United Kingdom

Dr. Michael Waldichuk
West Vancouver Laboratory
Dept. of Fisheries and Oceans
West Vancouver, B.C. V7V 1N6
Canada

Dr. C. J. M. Wolff
Shell Research B.V.
Koninklijke/Shell-Laboratorium
Postbus 3003
NL-1003 AA Amsterdam
The Netherlands

Hydrology

Reimer Herrmann

Lehrstuhl für Hydrologie, Universität Bayreuth
D-8580 Bayreuth, Federal Republic of Germany

Summary

This short article focusses on those general principles of the occurrence and movement of water which are important elements for the understanding of pollutant behaviour within the hydrological cycle. A defining introduction is followed by discussion of the components of the hydrologic cycle. The fluid dynamics of infiltration, saturated and unsaturated soil moisture flow, ground-water motion, runoff from snowpack, overland flow, and channel flow are developed. Erosion and sediment transport are treated in a further chapter. There follows a discussion of primary mechanisms of flow of water through plants and evaporation from soils.

The core of hydrological studies, the rainfall-runoff relationship and the water balance, are introduced as a basis of catchment modeling. Finally, the mass transport of chemicals within the hydrological cycle is examined.

Introduction

The sciences dealing with water can be divided [32] into four core subjects (Fig. 1). Within the geo-sciences these are oceanography and hydrology with physics, chemistry and mathematics as fundamental disciplines and biology and geo-sciences as complementary sciences.

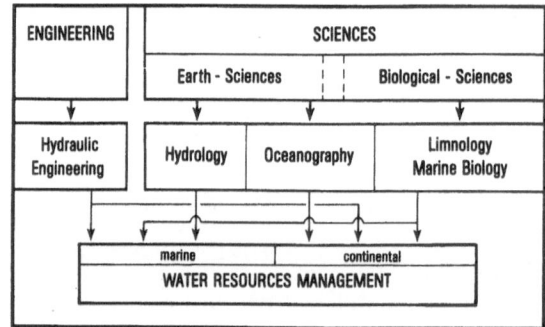

Fig. 1. Core subjects of the sciences of water (after De Haar [32])

De Haar [32] defines hydrology and oceanography as the sciences dealing with the behaviour of water above, on and below the earth's surface in respect to its distribution in time and space, its circulation and its physical, chemical and biological caused properties and effects. Thus, hydrology is the science of water on the continents and oceanography is the science of water in the seas.

Limnology and marine biology are biological sciences and are investigating the waters as biotops, thus they are sciences of the life in water and water as part of the biosphere. Limnology and hydrology are complementary sciences which share the same fundamental disciplines.

Hydraulics [87] belongs to the engineering sciences and serves as a base of hydraulic engineering [184] with physics and mathematics as fundamental disciplines. All four core subjects, hydrology, oceanography, limnology and marine biology as well as hydraulic engineering are fundamental subjects of water-resources management (Fig. 1). Economics, ecology, geo-sciences, regional planning, etc., are complementary to water resources management. Water resources management deals with the protection and the economically and ecologically tenable use of the earth's water resources.

Examples of more extensive textbooks on hydrology, include those by Eagleson [45], Raudkivi [141], Dracos [41] and Dyck and Peschke [44]. Particularly detailed and applied treatments of hydrology are to be found in Chow [20], Dyck [43] and Shaw [159].

The Components of the Hydrological Cycle

In order to give an introduction into the hydrological cycle, it may be convenient to confine it to the hillslope hydrological cycle [18], because nearly all important components may be included (Fig. 2).

The water balance and the transport processes of water within the hillslope hydrological cycle are predominantly influenced by the input of water as precipitation, primarily in form of rainfall and snowfall. Further, persistant foggy weather may have a relatively small but additive effect on the water balance of densely-forested river basins. The subsequent flow of water within the physical components of the hillslope (and finally river basins) is predominantly influenced by the droplet sizes and intensities of rainfall as well as of the nature of the snow flakes and the melting process.

Rainfall storage on vegetation, however, short, is termed interception. The part

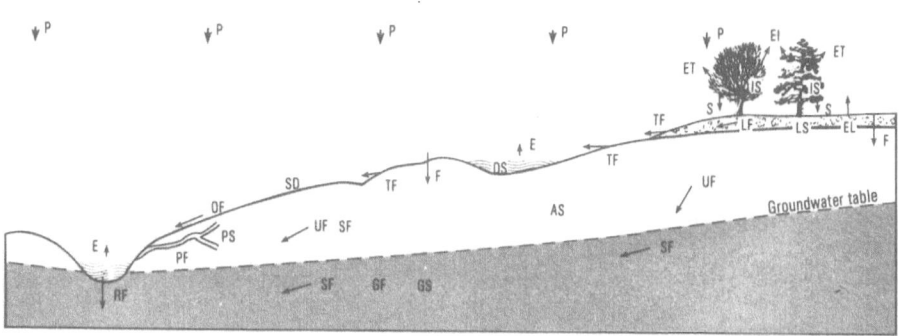

Fig. 2. Components of the hillslope hydrological cycle

P	= Precipitation (rain, snow, fog)		UF	= Unsaturated flow
E	= Evaporation		SF	= Saturated flow
ET	= Evapotranspiration		PF	= Pipe flow
EI	= Vegetation interception loss		F	= Infiltration
EL	= Litter interception loss		IS	= Vegetation interception storage
S	= Stemflow and throughfall		LS	= Litter storage
LF	= Litter flow		DS	= Depression storage
OF	= Overland flow		SD	= Surface detention
GF	= Ground-water flow		PS	= Pipe storage
RF	= River (lake) runoff		AS	= Aerated zone storage
TF	= Throughflow		GS	= Ground-water storage

of precipitation which is held back by plants and which evaporates from the plant surface or litter is called interception loss (Fig. 2). As this part of precipitation does not reach the soil, it does not contribute to infiltration into soil or to surface runoff. The extent of interception depends on the nature of the surface of the vegetation and litter (e.g. needles, leaves, age, phenology) and on meteorological influences (e.g. evaporation, rainfall intensity and temporal distribution, snowfall characteristics). Even if the interception is lost for water available to plants and to ground- and surface-water supply, it, however, keeps the water loss by transpiration of the plants down as during interception the gradient of vapour pressure across the leaves is small. That part of the water which reaches the land surface and turns directly into vapour from bare soils, from rivers and lakes, and from wetted litter and plants is termed evaporation (Fig. 2). Physiologically controlled release of water by plants is called transpiration. Actually, evaporation and transpiration both are influenced by the same physical laws, including diffusion, dynamic and thermal turbulence and changes of phase. Since it is often difficult to distinguish between these two phenomena, the combined process is called evapotranspiration. The amount of evapotranspiration loss depends to a large degree on the effective water and energy supply.

The infiltration of precipitation into the soils can be regarded as a primary control of the amount of surface runoff and depression storage on the one hand and the amount of ground-water and soil-water recharge on the other hand. This conveyance of precipitation through the soil surface into the pores of the deeper soil depends on the hydraulic conductivity and the gradient of soil suction in the soil profile. Thus, at the beginning of precipitation, a steep moisture gradient commonly occurs at the wetting front. As this wetting front proceeds downward into the soil profile, the gradient of suction will decrease with increasing water availability. Thus, high infiltration intensities at the beginning of a rainfall event (or melting snow) will be followed later on by lower infiltration intensities controlled principally by the saturated hydraulic conductivity and the gravitational potential.

If the rainfall intensity is greater than the infiltration intensity, part of the rainfall will be collected and stored temporarily in surface depressions (Fig. 2). Only if the water overflows these surface depressions will overland flow occur. Depression storage, thus, is defined as that water which fills depressions during or shortly after a rainfall event which does not run off and will subsequently evaporate or infiltrate [83]. In addition to depression storage, that part of rainfall on melting snow which flows downslope is called surface detention [83]; subsequently, the combined surface detention and depression storage form the surface storage.

Chorley [18] has suggested that, although overland flow is quite common where the soils are thin and vegetation is sparse, little overland flow occurs on slopes with dense vegetation and thick soil cover, even in intense storms. Often during a storm, only localized parts of a river basin will produce overland flow, because rainfall intensity varies considerably with time and space. Furthermore, the antecedent moisture conditions which control the infiltration process vary over the river basin as well. The simple Horton model of overland flow based on rainfall intensity in excess of the infiltration rate is further complicated by the fact that water which has infiltrated already and then, because of soil hydraulic conditions, moves only a short distance within the soil before seeping out into surface depressions and becoming overflow again. This phenomenon is known as interflow or through flow.

As a consequence, it seems to be difficult [18] to observe the classical model of overflow beginning with a thin layer of water flowing downslope which then accumulates into surface depressions, which in turn overflow and fill downslope gullies and stream channels.

In the aerated zone above the ground-water, most of the time the flow of moisture is unsaturated and is governed by suction gradients of the soil, gravity forces and moisture, depending unsaturated hydraulic conductivities. Thus, the resultant moisture flow may be directed towards the roots, the surface or towards the ground-water, depending upon the three-dimensional pattern of soil-moisture potentials. However, during storms, the flow within the aerated zone may, at least for parts of the river basin and over the short term, become saturated and thus it will be directed into the soil profile by gravity alone. Commonly, water begins to flow downslope within the aerated zone under unsaturated conditions [97]. When the soil, especially above horizons with low permeability, becomes saturated as a result of infiltration by storms or by snow melt, significant throughflow to gullies and stream channels will occur [187].

Little is known of the contribution of throughflow by concentrated flow in pipes and fissures [60]. Another part of the infiltrated water will seep to the ground-water. Within the aquifer, the ground-water flows in accordance with the hydraulic gradients

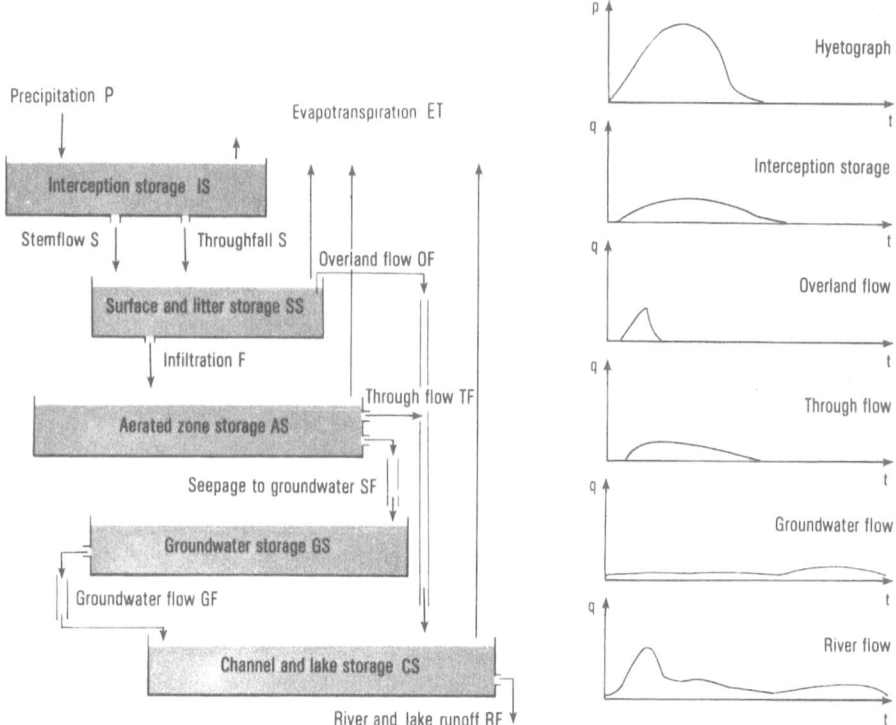

Fig. 3. Outline of water-balance components and hydrographs in a river basin

and hydraulic conductivity. This water, along with part of the throughflow, contributes to the base flow of receiving streams.

When the throughflow, overland flow or ground-water enters the lakes or rivers, part or – in the case of arid regions – all of it may evaporate. The remainder will flow in the rivers eventually into the oceans. Pulled by gravity, the water flows along the slope of the river with the counteracting resistance of channel banks and stream bed. The flow in rivers varies with time and in space; the flow velocity is different from one point in the river to the other and might be unsteady at these points. Further, there is an interconnection between flow, erosion and sedimentation [42].

The regional water balance of a river basin, as opposed to the world water balance [5], is an open system. This open system can be described as a sequence of conveyor segments and storage reservoirs which represent the components of the hydrological cycle.

Simplifying the hydrological processes outlined in Fig. 2 further, part of the precipitation inputs passes the vegetation storage, the litter and aerated zone storage and the ground-water storage and a part reaches the receiving river or lake (Fig. 3). The storage reservoirs tend to dampen the input hydrograph; whereas, the conveyor segment acts as a fictitious channel to translate the water without changing its hydrograph and thus causes only a phase shift. The inputs and outputs of each storage reservoir and the rate of change of storage can be used to write the following water balance equation:

$$P - (OF + TF + GF) - ET = \pm (IS + AS + GS + CS)$$

P   = Precipitation
OF = Overland flow $\left.\right\}$ Runoff
TF = Throughflow
GF = Ground-water flow
ET = Evapotranspiration
IS   = Rate of change of interception storage
AS = Rate of change of aerated zone storage
GS = Rate of change of ground-water storage
CS = Rate of change of channel and lake storage

Infiltration

A detailed introduction into the infiltration process is provided by Hillel [75]. He defines infiltration as the process of water entry from the ground surface into the soil. The flow is generally directed downward when originating from rainfall or surface irrigation. However, water may also ascend from horizons with a higher water content from below. The infiltration may vary temporarily and spatially depending on rainfall variation and soil structure. Philip [135] and Swartzendruber and Hillel [170] have reviewed comprehensively the physics of infiltration.

Hillel [75] defines the infiltration rate (i, LT^{-1}) as the volume flux of water flowing into the soil profile per unit of surface area. Under the conditions of atmospheric pressure and water freely available at the soil surface, he [74] calls this property infiltrability.

If the rate of water availability by precipitation or irrigation is smaller than the infiltrability, the process of infiltration is mainly controlled by the water availability. However, if the water-supply rate exceeds the infiltrability, the infiltration process is predominantly controlled by the physical properties of the soil profile. In this case, some water is ponded and may or may not result in the overland flow (Fig. 4).

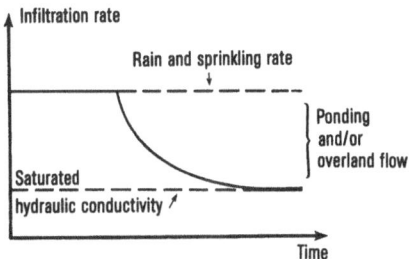

Fig. 4. Dependence of infiltration upon time (after Hillel [75])

The spatial and temporal varying physical properties of the soil [186] which lead to a spatial and temporal variation of infiltrability result from variation in initial soil-moisture and suction, soil structure and texture and the sequence of soil horizons. At the beginning of a rainfall event after a longer dry spell, the infiltrability is highest but with time the infiltrability decreases asymptotically to a value which is called by Hillel [75] steady-state infiltrability.

The decrease of infiltrability from an initial high rate to the lower steady-state infiltrability is caused by a decrease of the matric suction gradient due to the wetting of the soil profile. Further, swelling of clay minerals, entrapment of soil air, deterioration of the soil structure also contribute to a decrease of the infiltrability. In case of a stable and homogeneous soil profile with downward flow, the steady-state infiltrability equals the saturated hydraulic conductivity.

During infiltration of rain into a homogeneous soil, three zones of different matric suction may be observed (Fig. 5): Below a zone of full saturation which is only a few millimeters deep follows a transmission zone with low gradients of soil-moisture and matric suction. Below this zone, the wetting zone is found with steep gradients of both soil-moisture and matric suction. Within the upper two zones, the infiltration is predominantly controlled by gravity and the saturated hydraulic conductivity with only a small contribution of matric suction in the transmission zone; whereas, in the lower zone the flow is first of all controlled by matric suction and the unsaturated hydraulic conductivity.

Taking into consideration the great variability of infiltrability Table 1 gives a mean range of steady infiltration rates, which may be considerably higher (in dry soils with crevices) or lower (in soils with crusts). These data were compiled from Musgrave [120] and Hillel [75].

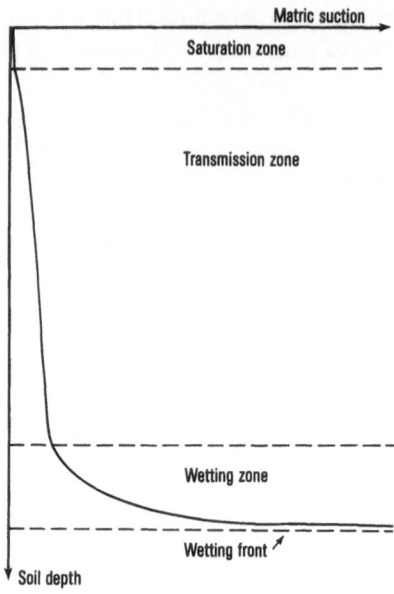

Fig. 5. Matric suction during infiltration except for special conditions (e.g. air entrapment or non-homogeneous soil profiles); the various zones show gradual transitions between each other

Table 1. Steady infiltration rates

Soil type	Steady infiltration rate (mm h^{-1})
Sands	> 20
Loess, silty soils with a good structure	10–20
Loam and sandy loam	5–10
Clayey loam and loamy clays	1–5
Sodic clayey soils	< 1

Mathematical Models of Infiltration

Swartzendruber and Hillel [170] have extensively discussed the useful infiltration models (e.g.) of Green and Ampt [59], Kostiakov [102], Horton [84], Philip [132] and Holtan [81]. Philip [135] discusses the infiltration process in great detail and shows that the infiltration rate can be described by a two-parameter equation:

$$i(t) = \frac{1}{2} st^{-1/2} + A, \qquad \text{where,}$$

$i(t)$ = infiltration rate
s = coefficient called sorptivity
t = time
A = coefficient

For a large time period t, A becomes the hydraulic conductivity K in the transmission zone, as in this case i = K. This equation is an approximation to the equation for unsaturated vertical flow (s. chapter 'Flow of Water in Unsaturated Soils') and is only valid for ponding water and not for rainfall infiltration.

As in nature, however, water generally infiltrates into highly nonuniform soils with respect to initial soil-moisture, texture and structure, more realistic theories of infiltration than those mentioned above had to be applied. These theories take into consideration a variation of the hydraulic conductivity K with soil-moisture content and depth as well as the variation of the matric suction with soil-moisture content and depth. Further, problems with different soil layers can be handled [e.g. 80, 79].

For example, a fine textured soil horizon with low hydraulic conductivity which lies underneath a coarse-textured upper soil horizon with higher hydraulic conductivity controls the infiltration process. This is because the overall infiltration rate will be that of the lower horizon when the wetting front starts to penetrate it. It is only an apparent contradiction that in the opposite case of a fine textured soil horizon above a coarse textured soil horizon, the infiltration is again controlled by the lower horizon: Initially, the infiltration is controlled by the fine-textured upper horizon, but when the wetting front reaches the interface to the lower coarse soil horizon, the infiltration rate may drop. This can be explained by the high suction at the wetting front which in turn is accompanied with a low hydraulic conductivity which does not allow a significant flux of soil-water into the large pores of the coarse soil horizon underneath. Since the low infiltration rate of the upper fine-textured soil horizon hinders the formation of saturated hydraulic conductivity in the lower horizon, sand and gravel beds under fine textured soil horizons may actually retard the infiltration.

The special case of infiltration into a crust-topped soil is treated by Hillel and Gardner [78]. Hill and Parlange [73], Raats [139] and Philip [136] have studied unstable flows in porous media. Especially at the interface between a fine-textured upper layer and a coarse-textured lower layer, the wetting front often does not move downward in an integrated front but rather protrudes downward in a fingerlike form. These authors explain this phenomenon, first of all, with water-repellent soil layers, air entrapment and increasing hydraulic conductivity with depth. By means of this unstable flow, solute-laden water might surpass the soil's storage and filter capacity and eventually pollute the ground-water. The infiltration of solute-laden water has been studied by Elrick et al. [47].

After the infiltration of water into a soil on other porous medium, a redistribution process begins which is controlled by hydraulic gradients and hydraulic conductivities. The hydraulic gradients and the (unsaturated) hydraulic conductivities are controlled by evaporation to the atmosphere, height of the ground-water table, initial potential distribution and root distribution, among other factors [76].

Soil-Moisture Content and Soil-Water Potential

The spatial variation of soil-moisture per unit mass or volume of soil and the potential energy of the water are important factors which cause the flow of water within the soils and out of the soils into plants, ground-water and to the land surface. The soil-moisture content can be described as either the mass ratio of mass of water M_w to mass of dry soil M_s

$$w = M_w M_s^{-1}$$

or the volume ratio of volume of water v_w to total soil volume v_t

$$\theta = v_w v_t^{-1} \ .$$

Both ratios w and θ are dimensionless. Furthermore, a ratio of $M_w v_t^{-1}$ (dim ML^{-3}) is widely used because of practical reasons. The soil-moisture content is measured by weighing, drying and again weighing, by electrical resistance, neutron scattering and other methods [10, 61, 64, 76].

The total potential of soil-water is defined [2] as "as the amount of work that must be done per unit quantity of pure water in order to transport reversibly and isothermally an infinitesimal quantity of water from a pool of pure water at a specific elevation at atmospheric pressure to the soil-water (at the point under consideration)". As several force fields (e.g. resulting from solutes in the soil-water, from the attracting forces of the soil particles' surface and from the gravitation) act on the soil-water, the total potential φ_t can be separated into three to these force fields related potentials:

$$\varphi_t = \varphi_o + \varphi_p + \varphi_g$$

φ_o = osmotic potential
φ_p = the pressure or matric potential
φ_g = gravitational potential
Other, generally not so important, potentials are not considered. The soil-water potentials can be defined as energy per unit mass (dim $\varphi = L^2 T^{-2}$) or as energy per unit volume (dim $\varphi' = L^{-1} T^{-2} M$) which is a dimension of pressure. The two forms of potentials are related to each other as follows:

$$\varphi' = \varphi \varrho_w$$

ϱ_w = density of water
The gravitational potential is positively measured in terms of relative elevation above

a reference level below the soil profile. If, however, the soil surface is used as the reference level, then the gravitational potential is negative at all points below that reference.

The gravitational potential φ_g is defined as follows:

$$\varphi_g = gz$$

g = gravitational constant
z = elevation

In soils, the pressure potential can exhibit two forms: The pressure potential is positive below a free-water surface (generally ground-water) and is termed submergence potential [76]. However, when the water is at pressure lower than atmospheric pressure, the potential is negative. This potential, when water is under suction, thus is called matric or (in older texts) capillary potential.

The submergence potential ($\varphi_{p,s}$) is defined as follows:

$$\varphi_{p,s} = gh$$

h = depth below the free-water surface

The matric potential can be obtained from measurements of soil suction

$$\varphi_{p,m} = -S\varrho_w^{-1}$$

S = suction
ϱ_w = density of water

The osmotic effect comes into effect only when semipermeable membranes like in plant roots hinder the free flow of solutes and by lowering the vapour pressure of soil-water. Thus, the soil-water flow is generally not influenced by osmotic potential gradients.

In dilute soil solutions, the osmotic potential can be measured from the osmotic pressure π which, for its part, can be computed by the van't Hoff equation:

$$\varphi_o = -\pi\varrho_w^{-1}$$

The functional relationship between soil-moisture content and matric suction can be represented graphically by the soil-moisture characteristic curve (Fig. 6). This indicats that the soil-moisture characteristic curve is strongly dependent upon the soil texture: The finer the texture the greater is the suction which is exerted by the soil on the soil-water. Since the dominant greater pores in a sandy soil are easily drained at a low suction, this results in a steep slope at lower soil-moisture contents. On the other hand, the more homogeneous pore-size distribution of clayey soils causes the curve to change only slowly. Besides the effect of soil texture on soil-water retention, the structure (aggregation and disaggregation, compaction) also has an influence on the soil-moisture characteristic curve.

As the equilibrium soil-moisture for each applied suction is greater in drying soils than in wetting soils, each of these two states exhibits a distinct soil-moisture characteristic curve (Fig. 7). This behaviour, which is called hysteresis [134, 175] is caused mainly by irregularly shaped pores, by different contact angles in sorption and desorption states, entrapped air, and swelling and shrinking phenomena.

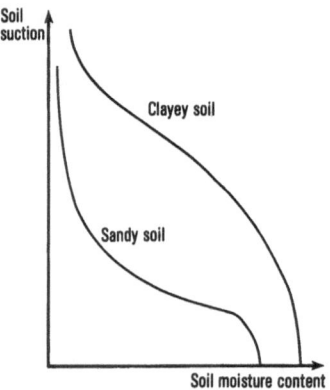

Fig. 6. The effect of soil texture on the soil-moisture characteristic curves

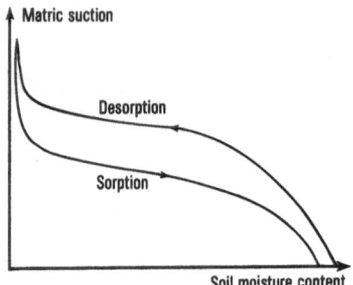

Fig. 7. Soil-moisture characteristic curves indicating the effect of hysteresis

Measurement of Soil-Moisture Potential

The matric potential can be measured indirectly by a tensiometer in the field or by tension plates and air-pressure extraction cells in the laboratory [144, 145]. The total soil-moisture potential can be determined from the equilibrium vapour pressure of soil-water by means of thermocouple psychrometer [28, 14].

Overland Flow

Overland flow is defined by Chorley [18] as "that part of streamflow which originates from rain which fails to infiltrate the mineral soil surface at any point as it flows over the land surface to channels". As rainfall as well as the infiltration rate varies in space and time, their difference causes an unsteady overland flow which varies also regionally. Depending on the velocity of flow and its depth, overland flow may be turbulent, laminar or one or the other at different parts of a hillslope. Furthermore, raindrops with high velocity may influence the flow.

In urban areas, a thin sheet of water flowing downward paved smooth surfaces develops with little change of depth perpendicular to the flow direction. However, on uneven natural hillslopes, often a lateral concentration of overland flow occurs which tends to erode and transport soil material.

Many time- and space-dependent factors, including rainfall intensity, infiltration rate, shape and steepness of the hillslope, its minor surface depressions and mounds and soil surface roughness, vary with time and/or space. Therefore, the resulting hydraulic characteristics will vary widely over space and time and subsequently are difficult to predict.

Hydraulics of Overland Flow

The basic hydrological concepts of overland flow were described by Horton in the thirties and early forties, culminating in his work on erosional development of streams and their drainage basins [85]. A widely-used standard textbook for surface-water hydraulics has been written by Chow [19]. Strelkoff [166, 167], and Rovey, Woolhiser and Smith [150] provide excellent operational introductions and Freeze [53] couples overland flow with subsurface flow and channel flow.

For spatially-varied unsteady flow-depth and flow can be described by the one-dimensional continuity equation with lateral inflow:

$$\frac{\partial h}{\partial t} + \frac{\partial (vh)}{\partial x} - r = 0$$

and the momentum equation for one-dimensional gradually varied, unsteady flow:

$$\frac{\partial v}{\partial t} + v\frac{\partial v}{\partial x} + g\frac{\partial h}{\partial x} + \frac{rv}{h} + g(S_f - S_0) = 0$$

where v is the local average velocity; h is the depth of flow; x is the distance of flow, t is time; g is the acceleration caused by gravity; S_f is the friction slope; S_o is the bed slope and r is the "inflow" by the effective rainfall along the slope constant over x as a step function ($r = r_1$ for $t < t_1$ and $r = 0$ for $t > t_1$).

These equations can be simplified by the kinematic-wave approximation to the momentum equation and solved by a finite difference method of solution [e.g. 150, 53].

Emmett [48], in his quantification of overland flow, estimates the unit discharge for turbulent flow for steady-state conditions by combining the continuity equation and the Manning equation:

$$q = \frac{1}{n} S^{1/2} D^{5/3} \quad \text{where}$$

q = unit discharge
S = slope
D = mean depth
n = Manning coefficient
This procedure is equally applied in open-channel flow calculations.
  And for laminar flow, the unit discharge is:

$$q = g \frac{SD^3}{3v} \quad \text{where}$$

g   = acceleration of gravity
v = kinematic viscosity
S, D = as defined above
Emmett [48] shows that, with increasing discharge, depth increases more rapidly with turbulent flow than with laminar flow.
  As a measure of the resistance to flow, the dimensionless Darcy-Weisbach friction factor f, is defined as:

$$f = \frac{8gDS}{v^2} \quad \text{where}$$

v = mean velocity.

Field and Laboratory Experiments

Intensive laboratory experiments with and without rainfall (see, for example, [164, 129, 128, 193, 195, 160, 48]) show an increase of depth with increase of roughness for a given discharge, with a greatest effect of roughness near the transition from laminar to turbulent flow [48]. The depth increases with decreasing slopes in a hyperbolic way. Falling rain disturbs the flow considerably; thus, this type of flow lies outside of laminar or turbulent flow, because it retards the flow and increases the depth for a given discharge [48]. High rainfall intensities retard the velocity at the upper slopes where the depth, surfaces velocity and thus momentum are small. Furthermore, the friction factor is enhanced by increasing rainfall intensities [160].
  The ratio of inertial forces to gravitational forces is called the dimensionless Froude's number; which is equally used in open channel flow (s. chapter 'Laminar and Turbulent Flow').

Emmett [48] observed that most flows in a smooth channel and all in a roughened channel were in subcritical ($F_r < 1$) flow. Over longer slopes, however, he also could observe supercritical flow ($F_r > 1$).

These laboratory experiments are compared with intensive field investigations by Emmett [48]. He found that the depth of flow in the field experiments were considerably greater than in the laboratory experiments. This is not unexpected, because an increased roughness in the laboratory experiments results in an increased depth of flow. Since the roughness in the case of field sites is a result of a combination from soil surface and vegetation resistance and from small geomorphic irregularities, the total retardance can be expected to be higher. Emmett [48] found friction factors (f) between 0.4 and 1.0 (mean 0.69) for field sites compared with a mean f = 0.48 for the rough-surface laboratory tests.

The higher f values are also a result of the characteristics of runoff (e.g. ponding). Emmett [48] proposes an average f value of 0.67 and no downslope change in roughness for estimation of hydraulic parameters. Also, the Froude's number was nearly constant in a downslope direction, with an average value of $F_r = 0.1$ indicating subcritical flow.

Erosion and Sediment Transport

Besides the contribution of overland flow to storm runoff, its contribution to solute transport and transport of sediments, nutrients and pollutants adsorbed on sediments are of great importance. The processes of erosion and deposition are explained in depth from hydraulic and sedimentological standpoints by Raudkivi [140], Graf [58], and Leeder [106]. De Ploey [33] has published a series of papers which enhance the progressing integration of field and laboratory studies in soil erosion, together with more theoretical models. Studies of Bagnold [3] and Emmett [48] seem to indicate that turbulence is not an essential requirement for sediment transport. Schumm [157] shows that an equilibrium exists between sediments and hillslopes in the way such that a decrease in slope angle on natural slopes is compensated by a decrease in roughness. Thus, the computed velocity was the same on steep and gentle slopes with an assumed constant depth downslope.

The widely used 'universal soil-loss equation' by Wischmeier and Smith [191] lacks the use of the kinetic energy of overland flow. Since its development, Bennett [7], Foster and Meyer [52], Williams [190], Kirkby [96] and Schmidt [156] described solutions closely related to hydrological overland-flow models. Intensive research in urbanized river basins [67, 105, 4] has shown that the covering of the surface by impervious material enhances greatly the overland-flow process. Furthermore, the infiltration of precipitation and thus the replenishment of the ground-water resources are restricted to the remaining pervious greenbelt, lawn, and park areas. The regional impacts of urban development in a river basin on the hydrological processes can be summarized as follows:

1) the direct runoff and the total volume of runoff are increased,
2) the peak flow of streams is generally increased,
3) the lag time of peak flow is shortened,
4) the impulse response exhibits a steep-rising limb,
5) low flow, which is fed principally by ground-water, decreases, and

6) pollutants are washed from roads and roofs and may cause a pronounced "first-flush" effect in the receiving streams.

A detailed introduction to urban stormwater modeling is provided by Overton and Meadows [127]. An elaborated hydrological model applicable to assessing effects of urban development was developed by Donigian et al. [39].

Ground-Water and Ground-Water Flow

Ground-water can be defined as that water which fills the pores and other hollow spaces of rocks and soils continuously and is subjected to gravity alone. At the water table (also called phreatic or free-water surface, Fig. 8), the pressure is equal to the

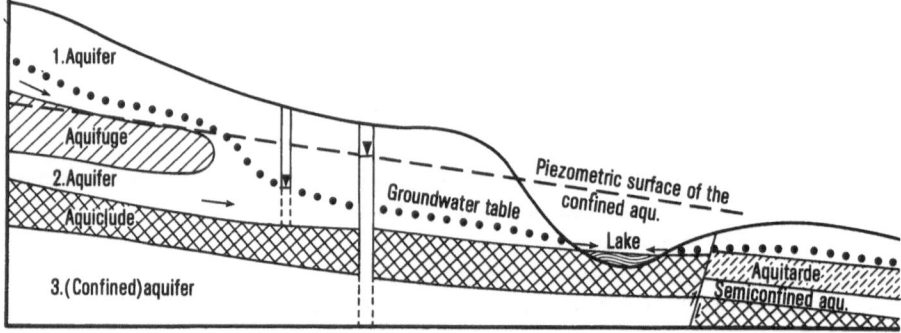

Fig. 8. Scheme of the occurrence of ground-water

atmospheric pressure. Those parts of rocks or soils which contain and allow a significant flow of ground-water are called aquifers, those, which do not, are defined as aquifuge. Confined aquifers are aquifers which are overlain by a relatively impermeable layer (i.e. an aquifuge) and when the water is under pressure, show an equilibrium height, also called piezometric level, which rises above the height of the aquifer. If a stratum is able to a small degree to transmit water vertically to an aquifer but not significantly horizontally, it is considered an aquitard. Aquifers sandwiched between one or two aquitards are defined as semiconfined or leaky. Aquicludes – in contrast to aquifers – contain and transmit only minor amounts of water. The terms aquitard, aquiclude and aquifuge are used for increasingly confining beds. However, the distinction between them, especially between aquitard and aquiclude, is not very precise.

Comprehensive treatments of hydrogeology and ground-water hydraulics are given in books by Todd [174], Remson et al. [143], Moser and Rauert [118], Brown et al. [15], Langguth and Voigt [104], Mattheß and Ubell [113], De Wiest [37], Verruijt [183], and Hálek and Švec. [62]. Examples of texts for ground-water quality and ground-water pollution are those by Mattheß [112] and Fried [54].

Flow of Confined Ground-Water

Darcy [30] showed that, at steady state, the rate of flow through an aquifer was proportional to the difference in hydraulic head and inversely proportional to the length
of flow (Fig. 9):

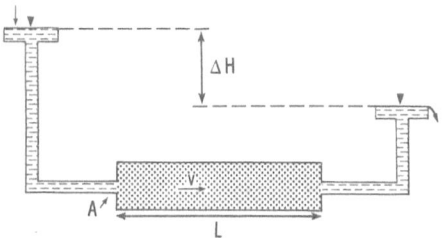

Fig. 9. Flow of water in a saturated soil column

$$Q = K \frac{\Delta H}{l} A \quad \text{where}$$

Q   = rate of flow
ΔH = difference in hydraulic head
l   = flow path
A   = cross section
K   = hydraulic conductivity
Darcy's law may be allowed for laminar flow.
    The apparent flow velocity or specific discharge can easily computed from:

$$v = QA^{-1}$$

The actual flow velocity, which takes into account the porosity n of the aquifer is
always greater than the apparent flow velocity:

$$v_a = vn^{-1}$$

n = porosity, n < 1
Thus, there exists a velocity distribution depending on the pore size distribution.
In saturated as well as in unsaturated flow the resulting hydrodynamic dispersion
strongly influences the mass transport of solutes.
    The hydraulic conductivity depends on the nature of the aquifer and the properties

of water (mass density ϱ and dynamic viscosity η). Taking these into account, the permeability k (dim k = L^2) can be derived:

$$k = \frac{K\eta}{\varrho g}$$

which depends only upon the nature of the aquifer. For most of the practical applica-tion η and ϱ have a low variation, and the hydraulic conductivity may be used.

Under conditions of steady flow, conservation of mass requires that the sum of changes of flow is zero. The resulting equation of continuity for flow in two dimensions, thus becomes:

$$\frac{\partial^2 h}{\partial x^2} + \frac{\partial^2 h}{\partial y^2} = 0$$

Flow of Unconfined Ground-Water

In an confined aquifer, the rate of flow can be estimated by computing the apparent velocity by Darcy's law and then multiplying it with the width and the height of the aquifer. However, in unconfined flow, one has to take into consideration that the cross-sectional area of flow diminishes (Fig. 10) along the flow gradient, due to the decrease of height with the water-table's slope. Integrating Darcy's equation for the flow rate (Fig. 10):

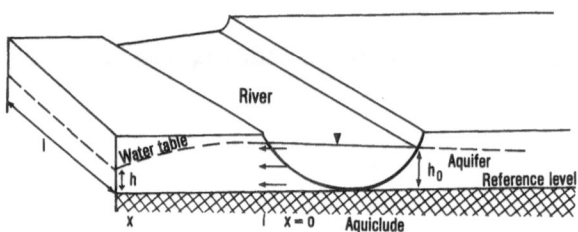

Fig. 10. Flow in an unconfined aquifer (after Bennett [6])

$$Q = -Klh \frac{dh}{dx},$$

one arrives at the solution:

$$h^2 = h_0^2 - \frac{2Qx}{Kl}$$

which shows the parabolic form of h. Thus, the parabolic slope of the hydraulic gradient counterbalances the decrease of the cross-sectional area [6].

Unsteady Flow

For conditions of unsteady flow in an aquifer, the sum of flow in the x- and y-direction equals the amount of storage within the aquifer. The storage coefficient is defined by the equation:

$$\Delta S = \mu \Delta h , \quad \text{where}$$

μ = effective porosity of the aquifer
Δh = change of ground-water level
Thus, using the continuity equation, the unsteady flow in an unconfined aquifer can be written as:

$$\frac{\partial^2 h^2}{\partial x^2} + \frac{\partial^2 h^2}{\partial y^2} = \frac{2\mu}{K} \frac{\partial h}{\partial t} .$$

In case of a confined aquifer, a change of piezometric head results in a release of water. This volume per unit area and unit change in piezometric head is called storativity S and can be equally used with the continuity equation to express confined unsteady flow [75].

The processes of confined and unconfined storage (also called specific yield) are different from each other. In the unconfined case, storage occurs at the water table; however, in the confined case, the storage is distributed through the entire aquifer. Also, the coefficient values differ significantly from each other:

$$0.01 < S_{unconf.} < 0.35 \quad \text{and} \quad 10^{-5} < S_{conf.} < 10^{-3} \text{ [6] .}$$

A wide range of direct and indirect methods are available for assessing ground-water flow rates and gradients and hence ground-water balance for a given system. Exemples include the application of tracers [118, 15], pumping tests [104], hydrograph analysis and water balance [113, 34].

As the ground-water is recharged by percolation through the unsaturated zone and otherwise above the ground-water table there develops a capillary fringe by soil suction, both zones have to be analyzed together. Furthermore, a hydraulic connection frequently exists between a ground-water system and an adjacent stream by which a rise of the water level of the river results in a rise of the ground-water table or, conversely, a rise of the ground-water table results in a steeper hydraulic gradient between river and ground-water which leads to an increased flow of ground-water into the river.

Because of the complicated regional pattern of aquifers, aquicludes and aquifuges and their hydrogeological properties as well as the geomorphology and the hydrological balance, the water table and the flow pattern of the various aquifers may exhibit a complicated pattern of regional gradients and temporal variability. Thus, the solution of practical tasks such as the flow in an aquifer, the flow towards a drainage system, the flow under dams or the flow to wells is complicated by a wide and often not well known spatial variation of the geohydrological properties. For practical applications the Dupuit-Forchheimer restrictions are often accepted which imply horizontal and constant flow throughout the whole aquifer and a small slope of the water table.

A detailed introduction into ground-water modeling is provided by Kisiel and Duckstein [98]. Well-documented two- and three-dimensional finite-difference models of ground-water flow are provided by Trescott et al. [177] and Trescott [176].

Flow of Water in Unsaturated Soils

As was shown in the chapter on ground-water flow, the driving force for saturated flow is a gradient of positive pressure potential. Unsaturated flow, on the other hand, is driven by a gradient of negative pressure potential which results from differences in matric suction within the soil. This gradient is highest at a wetting front where the gradient of the matric potential may exhibit a driving force of more than three orders of magnitude higher than the gradient of the gravitational potential.

This concept of unsaturated flow may be considerably complicated by cracks and macro-pores wherein water is able to infiltrate before a complete exchange with the fine matric has been reached. This accelerated infiltration is important for e.g. solute transport, leaching efficiency and the formation of through-flow.

The hydraulic conductivity in saturated flow is constant: whereas, in unsaturated flow it varies with the degree of occupation of the pores by the soil-water. Thus, when a dry soil is wetted, the cross-sectional area is increased and the hitherto long pathway of soil-water within a thin film around the pores will be shortened considerably. The variable composition of soil pores in different soils leads to a range of dependance of conductivity on suction (Fig. 11). In sandy soils, whose relative large pores are

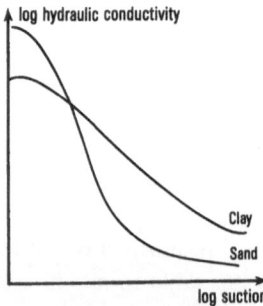

Fig. 11. Dependence of hydraulic conductivity on suction (after Hillel [76])

easily drained and thus the cross-sectional area is rapidly decreasing, the unsaturated conductivity is descending more steeply with a more even size distribution of fine pores than in clayey soils.

As pointed out by Hillel [76], this may lead to the observation that, in contrast to saturated-flow conditions, sandy soils may transmit less water than clayey soils under unsaturated conditions.

Since the soil-moisture content is a function of suction, it is also possible to express the unsaturated conductivity as a function of soil-moisture content.

Various empirical equations have been proposed in order to compute the unsaturated conductivity $K(\psi)$ or $K(\theta)$. The coefficients of these equations have to be derived by regression analysis, generally from laboratory tests. Some widely-used examples are:

$$K(\psi) = a(\psi^n + b)^{-1} \quad [56]$$

$$K(\theta) = K_s \left(\frac{\theta}{\theta_s}\right)^{2c+3} \quad [17] \quad \text{where}$$

ψ = suction head (measured by e.g. a tensiometer)
θ = volumetric soil-moisture content
s = subscript for saturated
a, b, c, n = empirical coefficients
Mualem [119] has developed a theory of unsaturated flow based on pore size distribution (s. also [181]).

According to Hillel [76], the dependence of conductivity upon soil-moisture content $K(\theta)$ seems to be affected by hysteresis to a much lesser extent than the dependence upon matric suction $K(\psi)$, which favours Campbell's [17] equation.

Darcy's flow equation can be written for unsaturated three-dimensional flow:

$$q = -K(\psi) \, \nabla(\psi - z)$$

$$\nabla = \frac{\partial}{\partial x} + \frac{\partial}{\partial y} + \frac{\partial}{\partial z} \quad \text{where}$$

q = three-dimensional flow
ψ = suction head
z = gravitational head.
   The continuity equation:

$$\frac{\partial \theta}{\partial t} = -\nabla \cdot q$$

and the Darcy equation can be combined to obtain the equation of three-dimensional unsteady and unsaturated flow [76]:

$$\frac{\partial \theta}{\partial t} = -\nabla \cdot [K(\psi) \nabla(\psi - z)]$$

This equation can be simplified for a horizontal one-dimensional system with $\nabla\psi \gg \nabla z$, in which case:

$$\frac{\partial \theta}{\partial t} = \frac{\partial}{\partial x} \left[K(\psi) \frac{\partial \psi}{\partial x} \right]$$

Measurement of Unsaturated Hydraulic Conductivity

Klute [99] has described methods for measurement of unsaturated conductivity in the laboratory. In principle, a soil sample is desorbed from its soil-water with tension plates or in a pressure chamber. Then, at discrete levels of pressure head and moisture content, a flow is initiated by imposing a hydraulic gradient across the sample.

Field methods are also available by which with a greater soil volume, the unsaturated conductivity can be measured [76]. These can be divided in three groups of methods: (1) sprinkling infiltration [196, 151, 142], (2) infiltration through an impeding layer [77, 12] and (3) internal drainage [57].

Movement of Water Vapour in Soils

Miller and Klute [114] assume that liquid flow controls the flow of water in moist, nearly isothermal soils. Only if a strong vertical temperature gradient occurs the driving force exists for a greater vapour flux. This flux can be described by the diffusion equation, because the vapour movement generally occurs by diffusion:

$$q_v = -D_v \frac{\partial \varrho_v}{\partial x}$$

D_v = diffusion coefficient in soils
ϱ_v = vapour concentration
q_v = vapour flux
Under consideration of relative small changes of vapour density with time compared with those of liquid water, which serves also as a source or sink for water vapour, Jackson [86] developed an equation for unsteady vapour transport:

$$\frac{\partial \theta}{\partial t} = \frac{\partial}{\partial x} \left(D_v \frac{\partial \varrho_v}{\partial \theta} \frac{\partial \theta}{\partial x} \right)$$

θ = volumetric moisture content

Particular elementary and lucid introductions into the simulation of transport processes in soils are to be found in De Wit and Van Keulen [38], Frissel and Reininger [55] and Hornung and Messing [82].

More detailed discussions of the behaviour of water in soils, which was outlined in the preceding sections on infiltration, soil-moisture content and soil-moisture potential and flow of water in saturated and unsaturated soils, are given in texts by Black [10], Rose [148], Groenevelt and Kijne [61], Hartge [65], Hillel [75] and Hillel [76].

Evaporation from Soils and Transpiration Through Plants

Water can leave the soil-plant system and escape into the atmosphere by direct eva- poration from the soil surface or from wetted plant-surfaces or from ponded water. If the water is taken up by the roots of the plants and evaporated through the stomata or the cuticles, then this process is termed transpiration, even though this may be in strict physical sense also be evaporation. Since it is often difficult to differentiate between these two processes, they are combined and named evapotranspiration.

Evaporation from Soils

In order to evaporate water from the soil (or wetted-plant surfaces), energy must be supplied by radiation and advection and a vapour-pressure gradient must be main- tained. These conditions are controlled by meteorological factors such as solar radia- tion, air temperature, wind velocity and vapour pressure. Thus, the meteorological factors control the maximum evaporation from free-water surfaces. This may be termed potential evaporation (or potential evapotranspiration where plants are also involved).

Penman [130] introduced a physical-based equation to compute the potential evaporation by means of the energy balance and aerodynamic moisture transport. This equation also can be extended upon vegetated areas by introducing a physio- logical resistance. A rigorous treatment of the evaporation and transpiration processes is given by Monteith [116, 117].

$$LE = \frac{\Delta \gamma^{-1} H_o + LE_x}{1 + \Delta \gamma^{-1}}$$

E = evaporation from a free water surface (mm d^{-1})
H_o = net radiation (cal cm^{-2} d^{-1})
E_x = isothermal evaporation (mm d^{-1})
Δ = slope of the saturation vapour pressure versus temperature curve at tempera- ture T (mm Hg $°C^{-1}$)
γ = psychrometric constant (0.485 mm Hg)
L = latent heat of evaporation of 0.1 cm^3 (= 59 cal)
The aerodynamic term E_x was empirically derived as follows:

$$E_x = 0.35 \, (0.5 + 0.54u_2) \, (e_s - e_2)$$

u_2 = wind velocity at 2 m (m s^{-1})
e_s = saturated vapour pressure (mm Hg)
e_2 = vapour pressure at 2 m (mm Hg)
A comprehensive review and practical applications of evaporation calculations are presented by Kijne [94].

Besides the two controlling factors – energy supply and vapour removal – the supply of water from the soil regulates the evaporation process.

A drying top soil provides a matric suction gradient which draws the water upward if there is sufficient soil-moisture available in the lower soil profile.

The steady flow in vertical upward direction can be described [133, 56] as follows:

$$q = K(\psi) \left(\frac{\psi}{dz} - 1 \right) \qquad \text{where}$$

q = flux-evaporation rate e
ψ = suction head
K = hydraulic conductivity
To obtain a semiempirical equation for evaporation e from soils, Gardner's [56] empirical equation for $K(\psi)$ (s. Chapter "Flow of Water in Unsaturated Soils") can be combined into the above equation [76]:

$$e = q = \frac{a}{\psi^n + b} \left(\frac{d\psi}{dz} - 1 \right)$$

The actual evaporation is controlled by the potential evaporation and the ability of the soil to transmit water of which the latter one controls the actual evaporation.

Hillel [76] points out that steady-state conditions of evaporation occur rarely, because of variability of potential evaporation and soil-moisture supply. For an initial and short period after infiltration, the soil can supply enough soil-water to meet the demand of potential evaporation. Then, after this weather-controlled initial period, evaporation falls below the potential evaporation rate because of the limited ability of the soil to transmit water to the soil surface or roots. This period is termed "profile-controlled". A dry soil which is able only to transmit water in the vapour phase is in a vapour-diffusion stage.

Transpiration Through Plants

In order to maintain a flow of water through the soil towards the roots, through the plant and out of the plant, it is necessary to keep up potential gradients within the soil-plant-atmosphere system and to supply sufficient soil-water.

The rate of water flow towards the roots is controlled by gradients of soil-water potentials, the relationship between the hydraulic conductivity and the soil-moisture or suction and the soil-moisture content. The uptake of water by the plant is regulated by the area of the soil-root interface, root depth and the internal water potential of the plant. Further, meteorological factors control the transpiration rate. If the transport of water across the soil/root interface falls below the transpiration rate, the resulting imbalance will over time result in a loss of turgidity and wilting of the plant. Rose [148] explains the relationship between the water potential and the different resistances acting upon transpiration within the soil-plant-atmosphere system (Fig. 12).

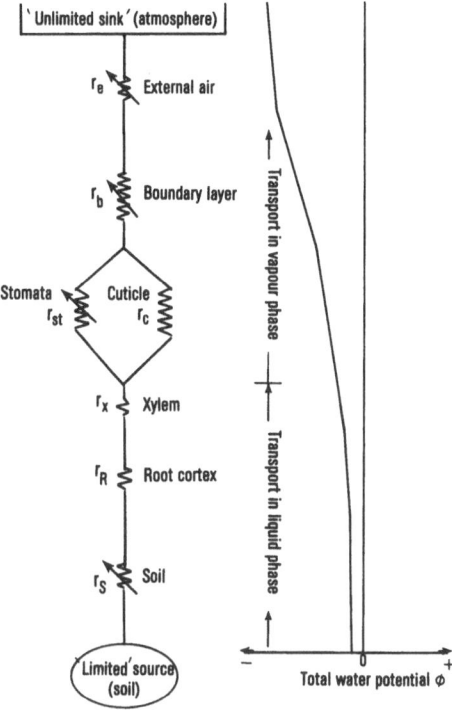

Fig. 12. Explanation of relationship between constant and variable resistances in the soil-plant-atmosphere system and total water potential (after Rose [148])

Following his concept of potential evaporation of 1948 [130], Penman in 1949 [131] coined the term potential transpiration. This is the evaporation of a well-watered stand of plants which is therefore controlled principally by the meteorological conditions.

Comprehensive treatments of the plant-water relationship are given by Slatyer [163] and Lange et al. [103]. A summary of related models is provided by Molz [115].

Water Balance of the System Soil-Plant-Atmosphere

Water balances usually are computed for river basins or for horizontally homogeneous unit areas of an ecotop, such as a unit area of a field. The water balance may be written as (s. Fig. 13):

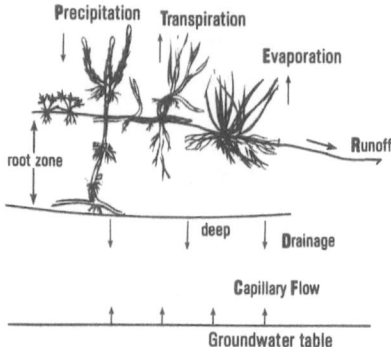

Fig. 13. Water-balance of the soil-plant-atmosphere system

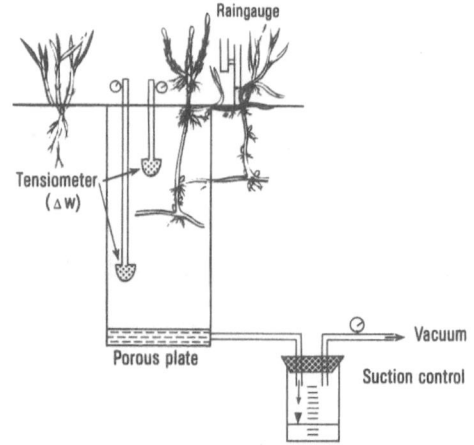

Fig. 14. Lysimeter with suction control and tensiometers

$$\Delta W_s + \Delta W_p = P + F_c - R - D - E - T$$

ΔW_s = change in water content in the root zone
ΔW_p = change in water content in the plants
P = precipitation and/or irrigation
F_c = capillary upward flow from the ground-water into the soil
R = runoff

D   = deep drainage out of the root zone
E   = evaporation
T   = transpiration

All terms are expressed in the dimension of volume of water per unit area which is equivalent to depth of water (dim $L^3 L^{-2} = L$) over the interval of time under consideration.

The most direct way of measuring the water balances is by means of lysimeters [149, 11]. Lysimeters (Fig. 14) are containers filled with soil and planted with the same vegetation as around it. They are designed to allow the measurement of deep drainage (e.g. by a suction plate) and the change of suction and hence water content (e.g. by tensiometers). Whenever the suction above the plate falls below a predetermined suction at the plate, percolation occurs and is measured.

As the change of plant-water content is relatively unsignificant as is also the runoff in irrigated fields, the most difficult to measure component of the water balance, namely the evapotranspiration (E + T) can be computed by the above given equation. The remaining component R can be obtained readily from rainfall data and irrigation measurements. The porous plate in the lysimeter (Fig. 14) allows no capillary upward flow, thus the term F_c can be omitted. If the total field water balance shall be investigated, however, then all components have to be measured, including the three-dimensional unsteady soil-water flow via tensiometers and soil-moisture probes [95].

Runoff from Snowpack

The knowledge of flow through snow melt is of increasing importance: rapidly melting snowpacks may cause damaging floods and snow cover can store and suddenly release pollutants enriched within the snow, causing steep pollutographs in the receiving streams [88, 23].

Colbeck [21] developed the theory of the movement of meltwater through a snowpack based on Darcy's law on unsaturated flow. These principles are extended and summarized in subsequent papers [22, 185] and applied to field measurements [138, 89, 90].

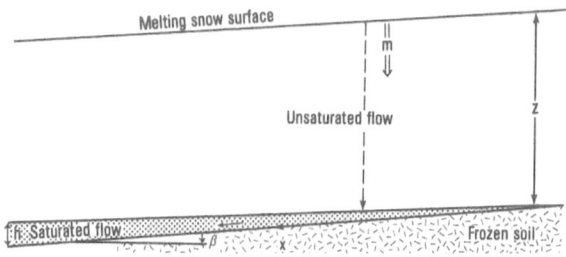

Fig. 15. Melting snow cover with unsaturated and saturated flow
m = flux rate at snow surface; z = depth of snow pack; h = height of saturated zone; β = angle of slope (after Price et al. [138])

    Movement of water through a snowpack is treated by Price et al. [138] in three
steps (Fig. 15). A heat-flow model is used to predict the melt rates. For an isothermal
snowpack at °C, the heat balance may be written:

$$H_m = H_r + H_c + H_e + H_p + H_g, \quad \text{where}$$

H_m = heat available for melting snow
H_r = net radiative heat flux
H_c = sensible heat flux
H_e = latent heat flux
H_p = heat gained from precipitation
H_g = ground heat flow

Melting at the surface causes a wave of meltwater to be released, which subsequently
moves downward through the unsaturated zone of the snowpack (Fig. 15). When
the water arrives at the soil it infiltrates or, if the soil is frozen, it flows along a thin
zone of saturated flow downslope. Following the procedure applied for flow in the
unsaturated soil zone, the equations for one-dimensional, vertical, downward, and
unsaturated flow and the equation of continuity are combined as follows:

$$\frac{\partial}{\partial z}\left(K\left(\frac{\partial \psi}{\partial z} + 1\right)\right) = \frac{\partial \theta}{\partial t}$$

Colbeck [21] assumed that for steady or decreasing flow, the suction term can be
neglected, which leads to the continuity equation for the speed of a flux wave [138]
through a snowpack:

$$\left(\frac{dz}{dt}\right)_m = \frac{n}{\varphi_e}\left(\frac{\varrho g k_u}{\mu}\right)^{1/n} m^{(n-1/n)} \quad \text{where}$$

n = an exponent in $k(S) = k_u S^n$
$k(S)$ = permeability of snow in the unsaturated zone to water at some fixed value of S
S = $(S_w - S_i)/(1 - S_i)$
S_w = water saturation, water volume per unit pore volume
S_i = irreducible saturation of the snowpack
k_u = permeability of the snow in the unsaturated zone
φ_e = effective porosity of the snowpack $\varphi(1 - S_i)$
φ = total porosity of the snowpack
ϱ = density of water
g = acceleration due to gravity
μ = viscosity of water
m = flux rate of water per unit area per unit time

This equation for unsaturated flow allows for the computation of the arrival of the meltwater at the bottom of the snowpack. The input I(x, t) can vary temporally and spatially. Thus, the continuity equation for saturated flow downslope for a unit width and a constant small angle β is:

$$I(x, t) = \frac{k_s \varrho g \beta}{\mu} \frac{\partial h}{\partial x} + \varphi \frac{\partial h}{\partial t}$$

k_s = permeability of the saturated zone
h = height of the saturated zone

A more detailed discussion of snow hydrology and glaciology is provided by Wilhelm [189]. Simplistic methods for snowmelt prediction relate daily meltwater flow to degree-days [24] or apply statistical analysis [197]. Both methods should be tested before use outside the area for which they were developed.

Channel Flow

Introduction

Flow in rivers is analyzed by the classical methods of open channel hydraulics. Extensive discussions of open channel hydraulics are given by Chow [19] and Streeter [165]. A more applied introduction of open channel flow is given by Jansen et al. [87] and of flood wave routing by Plate et al. [137]. A particularly usefull introduction to fluid mechanics is given by Douglas [40]. Leeder [106] and Tritton [178] have written texts that deal with fluid movements from a geosciences point of view. Results of recent research on channel processes including water, sediment and catchment controls have been provided by Schick [154]. Shen and Yen [161] provide a summary of advances of open channel hydraulics of the last twenty-five years.

Flow in rivers is generally spatially and temporally variable, which means it is commonly unsteady and non-uniform. For some practical considerations, the variations in time of velocity, pressure, density and temperature may be small enough to assume a quasi-steady or steady behaviour. If there is no longitudinal change of velocity, the flow is called uniform.

For water, the molecular viscosity does not vary with the rate of shear, and it is called a Newtonian fluid. Water-saturated muds, however, exhibit a change of the viscosity with changing shear and thus are named non-Newtonian fluids. In turbulent rivers, eddies can develop which cause a resistance to flow, called eddy viscosity, which is not constant for a fixed temperature.

To ease the mathematical treatment in hydrodynamics, fluids are assumed to be incompressible and with no resistance to flow. The theory of these inviscous or ideal fluids cannot explain many transport phenomena of great importance in rivers and canals, however.

In real fluids, a boundary layer is formed where the flow is retarded by viscous forces and a velocity gradient decreasing away from the channel bottom or the stream banks exists. Within the "free" stream flowing above the boundary layer, no velocity gradient exists.

Laminar and Turbulent Flow

In natural open channels, the flow is generally gradually-varied, turbulent and sub-critical. In such rivers, the cross-sectional velocity distribution does not change abruptly along the flow path. When viscous forces dominate over inertial forces, the flow is known as laminar flow, in the reversed case as turbulent flow. The two flow regimes usually are defined by the ratio of these two forces, the dimensionless Reynolds' number (R_e):

$$R_e = \frac{\varrho L \bar{u}}{\eta} = \frac{\bar{u} L}{v} \quad \text{where}$$

ϱ = fluid density
L = characteristic length, considered equal to the hydraulic radius R $\left(R = \dfrac{A}{P} \right)$
A = water area
P = wetted perimeter
$\bar{u}$ = mean flow velocity
η = dynamic viscosity
v = kinematic viscosity
A river flow is turbulent if R_e is large and laminar if R_e is small.

The ratio of the inertial forces to gravitational forces is known as Froude's number. For open channel flow:

$$F_r = \frac{\bar{u}}{(gh)^{0.5}} \quad \text{where}$$

$\bar{u}$ = mean velocity
g = acceleration of gravity

h = hydraulic depth $\left(h = \dfrac{A}{T} \right)$
A = water area
T = top width
If $F_r < 1$ or $\bar{u} < (gh)^{0.5}$, the flow is subcritical, which means that the gravity forces are relatively more important and the flow is tranquil. In the supercritical state, when $F_r > 1$ or $\bar{u} > (gh)^{0.5}$ and the inertial forces are dominant, the flow is torrential.

In the case of rivers exhibiting laminar flow, which, however, rarely occurs, the velocity distribution is assumed nearly parabolic (Fig. 16). The vicous shear stress τ is defined as the dynamic viscosity η times the velocity gradient $\dfrac{du}{dy}$:

$$\tau = \eta \frac{du}{dy}$$

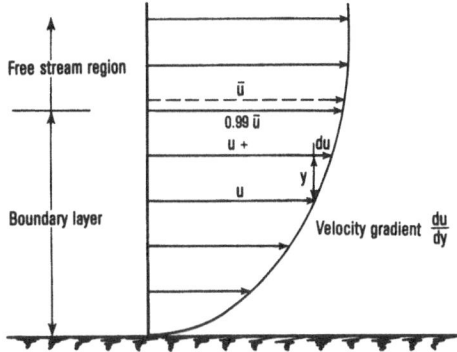

Fig. 16. Velocity distribution across a laminar flow over a solid bed

In turbulent flow, instantaneous fluctuations of velocity occur along all three dimensions, thus causing a variation of local stresses, which are known as Reynolds stresses of the form $\varrho\overline{u'v'}$, $\varrho\overline{u'w'}$ or $\varrho\overline{u'^2}$ originating from the fluctuating velocities u', v', w', respectively. These are additive to the viscous stress and are important for the transport of sediment and sediment-adsorbed pollutants. Thus, for turbulent two-dimensional flow, the shear stress τ should be adjusted as follows:

$$\tau_t = \eta \frac{du}{dy} - \varrho\overline{u'v'}$$

The distribution of total stress (τ), Reynolds stress ($-\varrho\overline{u'v'}$) and viscous stress $\eta(du/dy)$ across a turbulent boundary layer shows a thin layer which is called the viscous sublayer, wherein the stress is predominantly transfered by viscous forces (Fig. 17). Further away from the rigid boundary, the Reynolds stress dominates the contribution to the total stress.

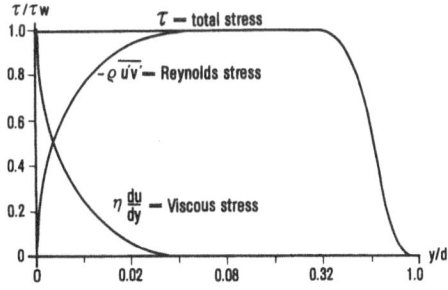

Fig. 17. Plot of total stress, viscous stress and Reynolds stress across a turbulent boundary layer. τ_w = wall stress; d = flow depth; y = height above bed (Re = $7 \cdot 10^4$); (after Tritton [178])

Velocity Distribution

Although over smooth boundaries with laminar flow a parabolic distribution of
point velocity is observed, in natural irregular channels with uniform turbulent
flow the roughness of the bottom will cause a logarithmic vertical velocity distribution
within the boundary layer. The velocity distribution over the cross-section is further
influenced by the shape of the cross-section and the curvature of the bends (Fig. 18).

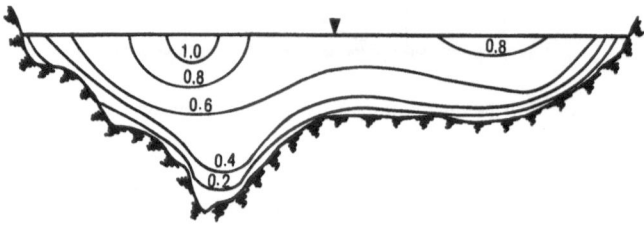

Fig. 18. Cross-sectional distribution of flow velocity (ms^{-1})

The velocity at any point in a turbulent flow over a solid surface is given by the
Prandtl-von Kármán universal-velocity-distribution law:

$$u = 2.5u_* \ln \frac{y}{c} \quad \text{where}$$

u = point velocity

$u_* = \left(\dfrac{\tau_0}{\varrho}\right)^{0.5}$ = shear velocity

τ_0 = shear stress on boundary
ϱ = fluid density
y = normal distance from the solid surface
c = constant for smooth surfaces depending on viscosity and shear velocity, varying
    with roughness

The derivation of the Prandtl-von Kármán law is outlined in Chow [19].

The effective height of the grains or sediment waves of a river bottom is called the
roughness height k. A surface is hydraulically smooth if all grains or sediment waves
are placed below the viscous sublayer. For this condition, Schlichting [155] found
experimentally that:

$$k < \frac{5\nu}{u_*}$$

ν = kinematic viscosity

The existence of smooth or rough surfaces in rivers play an important role for geochemical processes at the sediment-water interface.

If a channel widens along a short distance downslope or obstacles are placed into the river or if there are negative steps in flow direction (Fig. 19), pressure gradients will develop which may cause a boundary layer separation with a roller vortex. This phenomenon is discussed in more detail by Leeder [106].

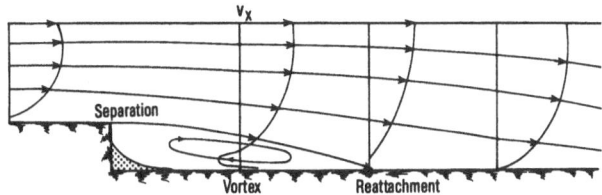

Fig. 19. Flow separation effect caused by a negative step (after Leeder [106])

Equations of Motions

The equations for flow in three-dimensions (Fig. 20) describe with a few approximations the conservation of mass and the conservation of momentum [87]:
Conservation of momentum

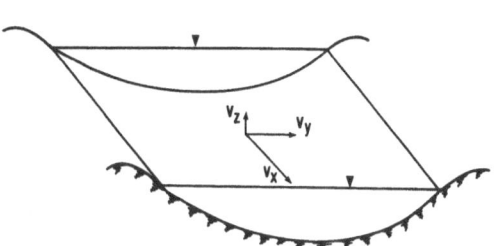

Fig. 20. Three-dimensional flow in a channel reach

$$\frac{\partial v_x}{\partial t} + \frac{\partial (v_x^2)}{\partial x} + \frac{\partial (v_x v_y)}{\partial y} + \frac{\partial (v_x v_z)}{\partial z} + g \frac{\partial h}{\partial x} - \frac{1}{\varrho} \frac{\partial \tau_{xz}}{\partial z} = 0$$

$$\frac{\partial v_y}{\partial t} + \frac{\partial (v_x v_y)}{\partial x} + \frac{\partial (v_y^2)}{\partial y} + \frac{\partial (v_y v_z)}{\partial z} + g \frac{\partial h}{\partial y} - \frac{1}{\varrho} \frac{\partial \tau_{zy}}{\partial z} = 0$$

Conservation of mass

$$\frac{\partial v_x}{\partial x} + \frac{\partial v_y}{\partial y} + \frac{\partial v_z}{\partial z} = 0$$

where
v_x, v_y, v_z = components of velocity in x, y, z directions
g   = acceleration due to gravity
h   = water level above a horizontal level
ϱ = density

$$\tau_{xz} = -\varrho \overline{v_x' v_z'} + \varrho v \, \frac{\partial \bar{v}_x}{\partial z}$$

$$\tau_{yz} = -\varrho \overline{v_y' v_z'} + \varrho v \, \frac{\partial \bar{v}_y}{\partial z}$$

τ = Reynolds stress + viscous shear stress = total stress
v = kinematic viscosity
The density of the fluid is assumed to be constant, which might not be the case in estuaries or rivers where salinity or suspended sediments cause variable densities.

Further, the Coriolis force has been omitted, because it shows an effect on flow only for rivers that are a few kilometers wide. The vertical acceleration is assumed to be very small compared to the acceleration due to gravity. Thus, the third equation of momentum is omitted.

Freeze [53] gives a particularly lucid development of the one-dimensional case of the equations of flow, which are called the "shallow-water" equations, and of their solution by means of the "single-step Lax-Wendroff" finite difference method.

For a more detailed discussion of finite-element techniques for fluid flow reference is made to Connor and Brebbia [25] and of the application of turbulence models in hydraulics to Rodi [147].

Estuaries

Cameron and Pritchard [16] define an estuary as a semi-enclosed coastal body of water which has a free connection with the open sea and within which sea water is measurably diluted with fresh water derived from land drainage. Thus, the open channel flow of the mouths of rivers discharging into the sea is influenced by horizontal gradients of salinity and, when flowing into seas with tidal action, by periodic tidal currents. Without tidal currents, a salt wedge would develop, causing the river water to flow above it out to sea. On account of tidal currents, sea and river water are partly mixed together, causing the circulation to be further changed.

A detailed discussion of the physical oceanography of estuaries is given by Officer [124]. Olausson and Cato [125] provide an excellent summary on the chemistry and biogeochemistry of estuaries. A detailed discussion of the modeling of estuarine hydrodynamics is given by Sündermann and Holz [169] and – together with water-quality aspects – by Orlob [126].

Various types of estuarine circulation are caused by the combined action of density currents, resulting from salinity gradients, river flow, tidal currents and estuarine bedforms. In many fjords and salt wedge estuaries, the estuarine circulation is governed

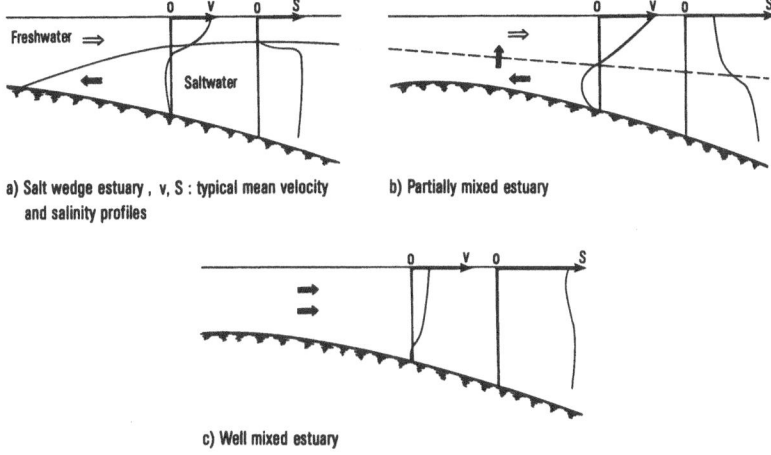

a) Salt wedge estuary, v, S : typical mean velocity
   and salinity profiles

b) Partially mixed estuary

c) Well mixed estuary

Fig. 21 a–c. Types of estuarine circulation

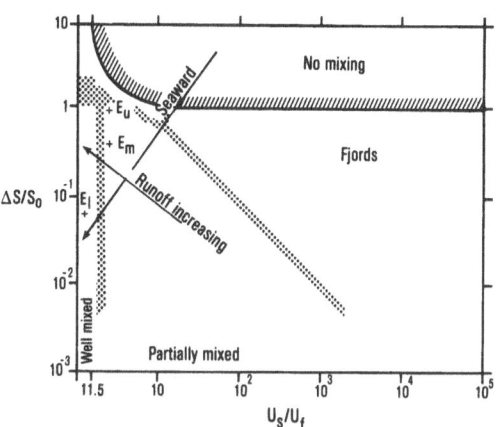

Fig. 22. Hansen and Rattray stratification-circulation diagram
E_l = lower Exe estuary; E_m = middle Exe estuary; E_u = upper Exe estuary; (after Herrmann and Hübner [71]; see text for further explanation)

by river flow. Such estuaries (e.g. the Mississippi) exhibit a strong stratification (Fig. 21 a).

If tidal currents mix the water vertically throughout the whole depth and if an upward residual current still exists below an downward residual current separated by a surface of no movement, the estuary is termed partially mixed (Fig. 21 b).

Estuaries become well mixed when the tidal action is very strong compared with river flow. These estuaries exhibit a small vertical variation of salinity. The mean vertical distribution of flow velocity shows no upward flow and thus the necessary upward compensating transport of salt is by means of diffusive processes (Fig. 21 c).

Transverse and lateral effects may further complicate the flow pattern considerably.

Hansen and Rattray [63] have introduced a stratification-circulation diagram by which it is possible to classify the estuaries (Fig. 22). The quotient $\partial S/S_0$, where ∂S is the difference of salinity between the bottom and the surface and S_0 the depth mean salinity, is a measure of stratification; whereas, the ratio u_s/u_f, where u_s is the mean tidal surface velocity and u_f the mean flow due to river discharge, is a measure of the estuarine hydrodynamics. The various types of estuaries inhibit different areas within the diagram. It should be mentioned that a single estuary can be assigned to different types because the lower end may be well mixed; whereas, the upper end is partially mixed.

Rainfall-Runoff Relationships

The investigation of the relationships between the rainfall over a river basin and the resulting runoff is a central element of hydrological research.

The analysis of this relationship depends very much upon the scales of time and space: the investigation of a flood in a medium-sized catchment has to take into account the depth-area-duration relationship of rainfalls [192, 159] and the spatial and/or temporal varying hydrological factors, such as infiltration rate, redistribution of soil-water, ground-water flow etc. With increasing time and area, these spatial and temporal variations are frequently leveled out. For a more detailed discussion of hydrometeorological influences reference is given by Wiesner [188].

A hydrograph of discharge as a function of time is formed by the combined rainfall and catchment characteristics (Fig. 23). The hydrograph can be divided into the volume of direct runoff (comprising overland flow and quick return flow ~ through flow) and base flow, which is contributed from ground-water.

That part of the rainfall which contributes to runoff is termed effective rainfall, the remaining part is lost through evapotranspiration. These parts can be distinguished by the loss curve.

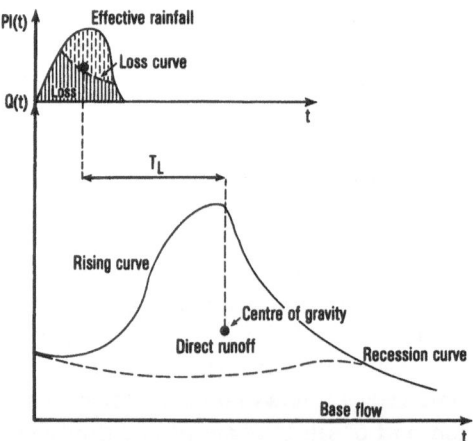

Fig. 23. Rainfall intensity, PI (t), and discharge hydrograph, Q (t)

When the initial direct runoff reaches a stream gage, the hydrograph rises even if the base flow is still falling. Later, the ground-water flow increases and causes the base flow to rise also. The lower receding part of the hydrograph is made up principally of water which had been travelling for a longer time through the soils and thus extends behind the time when the effective rainfall has already ceased.

The time T_L between the centres of gravity of the rainfall and discharge hydrograph is known as the lag and is a measure of the response time of a river basin.

A detailed discussion of hydrometrical methods to measure the rainfall runoff relationships can be found in Herschy [72].

Systems Approach to Derive Runoff Hydrographs from Hyetographs

The direct runoff from a unit uniform effective rainfall (e.g. 1 mm over the catchment) of specified duration (e.g. $\Delta t = 1$ h) is defined as unit hydrograph. Under the assumption of linearity and time invariance, the unit hydrograph may be convoluted with an effective rainfall hyetograph to compute a runoff hydrograph. If the unit hydrograph was derived from a specified effective rainfall of the duration Δt, which is of the same length as the time intervals of the effective rainfall hyetograph, any runoff hydrograph can be calculated by the convolution:

$$Q_j = \sum_{j=0}^{j-1} R_i U_{j-i} \quad \text{where}$$

Q_j = hydrograph ordinate at time $j \, \Delta t$
R_i = volume of water between $(i - 1) \, \Delta t$ and $i \, \Delta t$ of the hyetograph
U_{j-i} = Δt-unit hydrograph ordinate at the time $\Delta t(j - i)$.

If the unit uniform effective rainfall fell instantaneously over a catchment, that means that Δt becomes dt, because the duration approaches zero. The resulting limiting form of the Δt unit hydrograph is termed the instantaneous unit hydrograph. The discrete rainfall depth intervals R_i of effective rainfall are replaced by the continuous rainfall intensity function $R(\tau)$ times an infinitesimal time interval $d\tau$ (Fig. 24). The ordinates U_{j-i} are substituted by $h(t - \tau)$, which is the continuous form of the instantaneous unit hydrograph (IUH). An outline derivation of the theory is given by Singh [162]. For an introduction into practical computational procedures references are given to Dyck [43], Shaw [159] and Overton and Meadows [127]. Dyck and Peschke [44] evaluate the theory further by linear-time-invariant systems analysis, the terminology of which is added into Fig. 24. The convolution integral of the product of the effective rainfall function $R(\tau)$ and $h(t - \tau)$ replaces the discrete summation of $R_i U_{j-i}$ and gives the continuous hydrograph $Q(t)$ in place of Q_j:

$$Q(t) = \int_0^t R(\tau) \, h(t - \tau) \, d\tau \quad \text{where}$$

τ is the time variable of integration.

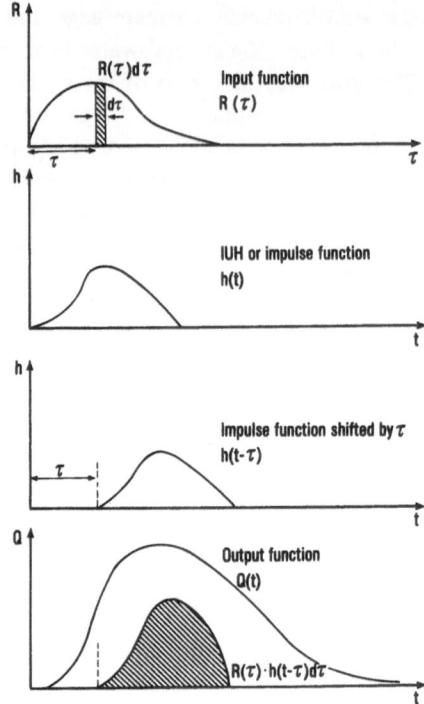

Fig. 24. Convolution of R (τ) with the impulse response function h (t)

Although basically the computation of the runoff hydrograph by means of a continuous rainfall and the IUH follows a "black box" concept which does afford previous time series of rainfall and related runoff to determine the IUH, mathematical models were developed for river basins with no runoff records using basin parameters to deduce the IUH or impulse response of the catchment [121].

Rainfall-Runoff Relationships Over Longer Periods

The investigation of the water balance on a monthly or yearly basis is commonly a focal point of water resources planning for a river basin. The water balance is dependent upon the climate, the vegetation, the soils, the hydrogeological structure, the relief and the degree of urbanization and industrialization.

On a monthly basis, a water-balance equation can be written as follows:

$$P = ET + R + \Delta S \quad \text{where}$$

P = precipitation
ET = evapotranspiration
R = runoff
ΔS = change in storage (both surface and subsurface)

Table 2. Water balance of the Eder at Müsse (1950–61) (mm)

	N	D	J	F	M	A	M	J	J	A	S	O	Σ
1. ET (%)	2	1	1	2	5	8	16	17	17	15	11	5	100
2. P	107	124	122	101	86	85	78	95	112	106	90	106	1212
3. R	77	113	132	95	106	62	31	29	41	51	65	75	877
4. L=P−R	30	11	−10	6	−20	23	47	66	71	55	25	31	335
5. ET	7	3	3	7	16	27	55	57	57	50	37	16	335
6. ΔS	23	8	−13	−1	−36	−4	−8	9	14	5	−12	15	0
7. ΣΔS	23	31	18	17	−19	−23	−31	−22	−8	−3	−15	0	

With the example of the mean monthly water balance for the Eder at Müsse (Rothaargebirge, Germany) (1951–1960) the procedure of the budget computation shall be introduced (Table 2, from Herrmann [68]).

The mean monthly runoff and the mean monthly precipitation were measured. The mean monthly percentage of ET, obtained from lysimeter measurements was taken from Keller [91] as indicated in row 1. First, the loss L is computed for each month and the year. Since the mean annual loss for a long time series equals the ET, the percentage monthly distribution of row 1 can be used to compute the monthly ET in row 5. The difference between monthly losses L and evapotranspirations ET (row 4–row 5) results in the change in storage ΔS. Finally, the storage capacity (= 62 mm) can be calculated from the range of the mass curve of ΔS (row 7).

A detailed discussion of the world water balance is given by Baumgartner and Reichel [5]. An excellent example of regional hydrology is given by Keller [92].

Catchment Modeling

One of the key hydrological problems has always been to estimate the flow of ungaged streams or to extend the available records of gaged streams. Unfortunately, when modeling these processes, one has to abstract from the real-world processes and to simplify the structure of the rainfall-runoff relationship, regarding the time-area-functions of rainfall, evapotranspiration, interception, infiltration, overland flow, soil-moisture redistribution, through flow and ground-water flow.

For introductory discussions of river-basin modeling, reference is made to DFG [35 and 36], Fleming [50], Overton and Meadows [127], and Linsley [108].

Catchment models – as other hydrological models too – can be classified into three groups, stochastic, deterministic and parametric models. They may further be subdivided into distributed or lumped models depending on their degree of spatial structure. In stochastic modeling, streamflow – or other hydrological functions – are assumed to be stationary stochastic processes. After the generating mechanisms are statistically approximated, it is possible to generate future flow sequences or synthetic flow sequences. Matalas and Wallis [111] have provided an excellent summary on the generation of synthetic flow sequences and their use has been discussed by Fiering and Jackson [49]. In contrast to the stochastic models which base on random processes, the deterministic models have a structure based upon physical laws, measures of initial and boundary conditions and inputs. Recently deterministic models have been developed which are also based on physical laws but who are supplied with statistical distributions of the parameters [29]. In parametric models, the model parameters are not necessarily physically defined. Thus, the parameters must be found by fitting the model to measured hydrological data by an optimization procedure of one form or another. Both parametric and deterministic models allow the prediction of hydrologic events in space and/or time; whereas, stochastic models produce probabilistic outputs.

In order to describe function and structure of such models, one each out the three groups of models – parametric, deterministic and stochastic – was selected solely from a didactic point of view.

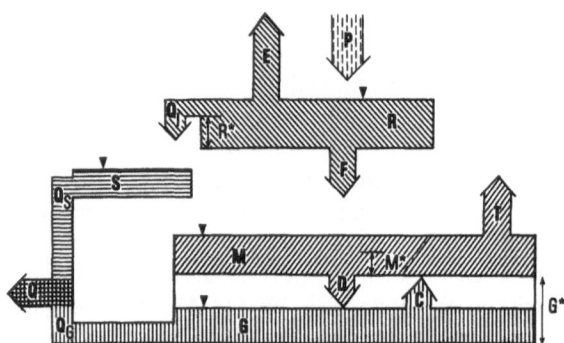

Fig. 25. Parametric model of Dawdy and O'Donnell [31], for explanation, see text

Dawdy and O'Donnell [31] proposed a parametric model which is based on four storage reservoirs (Fig. 25). The rain P falls into the surface-storage reservoir R where it is stored until a threshold R* is reached. From this reservoir, excess water F infiltrates down to the soil-moisture storage reservoir or runs (Q_r) into the channel storage reservoir S. The water from the surface storage reservoir R and the soil-moisture storage reservoir M can be depleted by evaporation E and transpiration T. The soil-moisture storage reservoir is drained by percolation D after reaching a threshold M*, into the ground-water storage reservoir G. Up to a ground-water threshold G*, water can rise vertically as unsaturated flow C into the soil-moisture storage reservoir. The (linear) reservoir runoffs Q_S and Q_G are the only outputs because no direct flows occur from the soil storage reservoir to the receiving river.

Dawdy and O'Donnell [31] used an optimization procedure to estimate the nine parameters used in the model, such that the difference between calculated runoff and measured runoff would be a minimum. These nine parameters belong to the set of equations comprising the mathematical model, and must be determined in advance for each catchment.

The three storage thresholds G*, M* and R*, the two linear reservoir constants K_S and K_G for the reservoirs S and G, the minimum rate of infiltration f_i and the maximum rates of infiltration f_a and the unsaturated vertical water flow C_a, and, finally, an exponent k in the infiltration equation must be optimized.

Crawford and Linsley [27] have probably developed the most elaborate deterministic hydrological simulation and prediction model which they called the Stanford

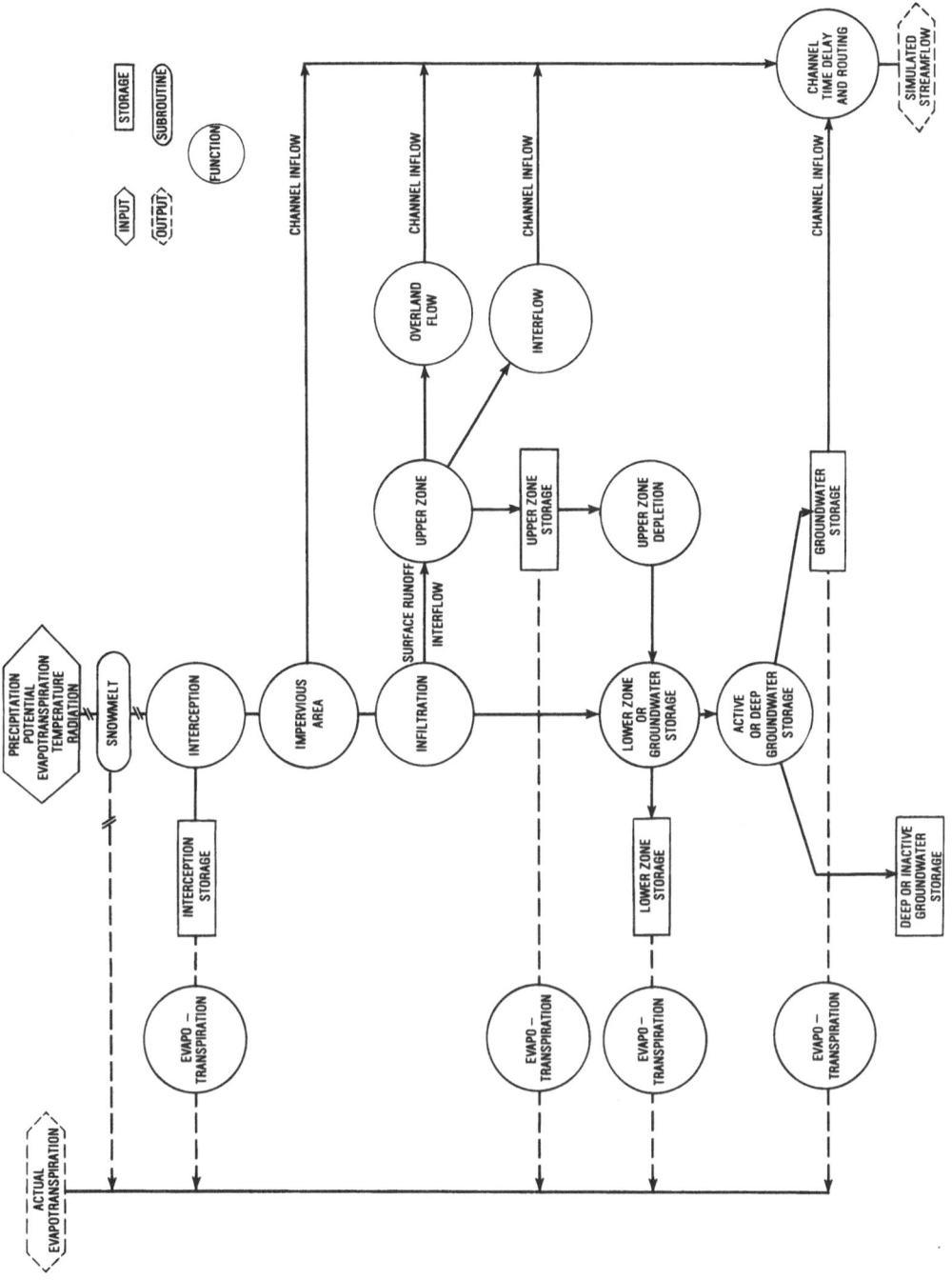

Fig. 26. Stanford watershed model IV, flowchart (after Crawford and Linsley [27])

Watershed Model IV (SWM IV). The latest and most completely tested model derived from the SWM IV is the Hydrocomp Simulation Program (HSP, Hydrocomp International). A flow diagram of SWM IV is shown in Fig. 26. Inputs required for the model include rainfall and potential evapotranspiration. In an initial step, the interception loss is simulated. There is an inflow into the receiving channel directly from impervious area, as overland flow and interflow after a short infiltration passage and finally as ground-water flow. The soil-moisture storage is the central point of the model; whereas, Philip's infiltration equation is probably the most important single function.

In order to elucidate the structure of the HSP a bit further, the function of the "upper-zone depletion" (s. Fig. 26) is described in some detail [108]. Water which is for a time stored in the upper-zone may infiltrate into the lower zone from depressions and vertical flow out of the upper-soil horizons. The rate of this percolation PERC is assumed to be a function of the relative moisture content of the upper and lower zones. Thus,

$$PERC = 0.003 \cdot INFILTRATION \cdot UZSN \left(\frac{UZS}{UZSN} - \frac{LZS}{LZSN} \right)^3$$

where INFILTRATION is the infiltration level parameter, UZSN and LZSN the nominal capacity of upper respectively lower-zone storage, UZS and LZS are the upper and the lower-zone storage respectively. Percolation from the upper-zone occurs only when UZS/UZSN > LZS/LZSN.

Fortunately, most of the parameters used in the functions describing the various hydrological processes can be easily obtained from watershed data. There are only four parameters which require a special calibration procedure [108].

One of the most widely applied stochastic models is the Markov flow model introduced by Thomas and Fiering [172]. Yevjevich [194] provides a detailed introduction on probabilities and statistics in hydrology. Markovian flow means that a given flow depends on the preceding flow and a random component. A particularly lucid and elementary introduction into the development of stochastic models are given by Fiering and Jackson [49]. They assume that the flow q_i is the sum of the mean μ, a proportion, given by the serial lag-one correlation coefficient ϱ, of the departure of the previous flow from its mean $\varrho(q_{i-1} - \mu)$ and a random component. If t_i is a Gaussian-distributed, serially-independent random variable with zero mean and standard deviation 1, then the random component $t_i \sigma \sqrt{(1 - \varrho^2)}$ is also Gaussian-distributed, serially independent, has zero mean and a variance $(\sigma^2 - 1\varrho^2)$. Then the Markov model:

$$q_i = \mu + \varrho(q_{i-1} - \mu) + t_i \sigma \sqrt{(1 - \varrho^2)}$$

generates Gaussian-distributed synthetic flows that preserve the mean, variance, and first-order serial correlation coefficient of the historical flow record.

In expanding this model, Fiering and Jackson [49] take into account seasonal variations and gamma or log-normally distributed flows. Further, strong influences of ground-water flow, which cause a long memory can be simulated by multilag models and finally use can be made of multi-site models where a regional inter-correlation of hydrological processes exists.

An extensive discussion of surface-water management models is given by Loucks [109]. For well-documented models of surface-water routing, channel routing, infiltration, hydrologic and water-control systems analysis, reference is made to Rovey et al. [150] and various HEC models [66].

Mass Transport within the Hydrological Cycle

The intention of this last chapter is to give a short guideline into an important and rapidly expanding field of theoretical and applied hydrology. Although research on sediment transport is already a classical disciplin of hydraulics and sedimentology [58, 140, 106] a comprehensive, and based on physical principles, treatment of transport of momentum, energy and mass was given by Bird et al. [8] not until 1960. A particularly lucid and elementary introduction into environmental movement of chemicals in air, water and soils is to be found in Thibodeaux [171].

The transport of a substance, e.g. a chemical, in space and over time within the hydrological cycle is governed by the movement within the flowing water, termed advection or convection, by various mixing processes comprising dispersion and by reactions like adsorption and desorption and different modifications of the chemical within the aquatic environment.

The combined advection-mixing and reaction processes can be represented by the following equation [126, 171]

$$\frac{\partial C}{\partial t} = -v_i \nabla \cdot C + \nabla \cdot (E_i \nabla C) + R$$

Change of concentration = Advection + Mixing + Reaction
C = concentration of the substance
v_i = vector velocity, i denotes the vector direction
E_i = effective diffusivity vector
R = reactions
The equation is written in this general vector form in order to show the three-dimensional behaviour of the advection and mixing terms.

v_i is the field velocity or advective velocity by which the substance is transported by the flow of water. However, if a chemical is carried in an adsorbed phase by suspended sediments (e.g. PAH on silt in a river) the advective velocity is that of the particles. In an analogous way the effective diffusivity for solid particles has to be used.

The effective diffusivity combines the molecular diffusivity D_m, the turbulent diffusivity ε_i and other random mixing processes E_{ai} including advective dispersion, wind induced mixing, tidal oscillation [126], i.e.

$$E_i = D_m + \varepsilon_i + E_{ai}.$$

Usually, in the general transport equation the movement terms have to be simplified for practical applications. For example, in streams the dominant mass transport occurs only in one direction, or in ground-water flow $E_i \sim D_m$.

The reaction term may comprise one or more processes and also reactions between the various chemicals. The degree to which such factors as advection or desorption from suspended sediments or the soil matrix, as photochemical processes, as redox reactions, as hydrolysis or as metabolic transformations and the like, may be treated, depends on their importance in the aquatic ecosystems under consideration [168, 179, 173, 101].

In order to illustrate the behaviour of substances within the hydrological cycle the modes of occurrence (dissolved and adsorbed) and the various ways of transport of 3.4-benzopyrene through an urban environment [69] are shown in Fig. 27.

Fig. 27. 3.4-Benzopyrene transport relationships in Bayreuth at a peak flow event (May 27, 1979) (Herrmann [69], with permission)

Examples for particularly detailed introductions into the behaviour in the aquatic environment of typical classes of substances are to be found for trace metals in Förstner and Müller [51] and Salomons and Förstner [152], for polycyclic aromatic hydrocarbons in Neff [123], for pesticides in Khan [93] and for organic micropollutants in Herrmann [70].

An advanced discussion of modeling and control of river quality is given by Rinaldi et al. [146]. A particularly lucid and elementary introduction into surface-water quality management models is to be found in Loucks [110]. For well documented computer simulation programs for water quality simulation of surface waters is reference made to NCASI [122] and Donigian et al. [39] and for pesticides and toxic substances to Crawford and Donigian [26] and Ambrose et al. [1]. A water quality simulation model for estuaries was developed by Leendertse and Gritton [107]. Extended discussions of case studies of water quality management models are to be found in Biswas [9]. Sanders et al. [153] have provided an excellent course in design of networks for monitoring water quality. For a more detailed discussion of ground-water quality modeling reference is made to Fried [54]. A well documented model of two-dimensional solute transport and dispersion in ground-water is provided by Konikow and Bredehoeft [100].

The understanding of the simultaneous transport of water and solutes under instationary saturated and unsaturated conditions is of importance for the optimal application of fertilizers as well the disposal of domestic and industrial wastes on fields and in landfills. Earlier studies on saturated-unsaturated transport have used finite difference techniques for solving the transport equations [13, 180], however, recently also finite element methods have been applied [158, 182]. An outline derivation of solute transport during infiltration is given by Elrick [46].

Acknowledgement: I wish to thank H. Rohdenburg and T. D. Steele for valuable advice. I also thank E. Misch and E. Schill for preparing the manuscript and the illustrations.

References

1. Ambrose, R. B., Hill, S., Mulkey, L. A.: User's manual for the chemical transport and fate model TOXIWASP, Version 1, EPA-600/3-83-005, U.S. Environmental Protection Agency, Athens 1983
2. Aslyng, H. C.: Int. Soc. Soil Sci. Bull. 23, 7 (1963)
3. Bagnold, R. A.: Proc. Inst. Civil Engrs. 6041, 174 (1955)
4. Balmér, P., Malmquist, P.-A., Sjöberg, A. (Eds.): Analysis and Design of Stormwater Systems. 3 vols., Proc. Third Int. Conf. Urban Storm Drainage, Chalmers Univ. Technology, Göteborg 1984
5. Baumgartner, A., Reichel, E.: Die Weltwasserbilanz, R. Oldenbourg, München 1975
6. Bennett, G. D.: Introduction to ground-water hydraulics, U.S. Geol. Survey, Arlington 1978
7. Bennett, J. P.: Water Resourc. Res. 10, 485 (1974)
8. Bird, R. B., Stewart, W. E., Lightfoot, E. N.: Transport Phenomena, Wiley, New York 1966[7]
9. Biswas, A. K. (Ed.): Models for water quality management, McGraw-Hill, New York 1981
10. Black, C. A.: Methods of Soil Analysis, part 1, Am. Soc. Agronomy, Madison 1965
11. Black, T. A., Thurtell, G. W., Tanner, C. B.: Soil Sci. Soc. Am. Proc. 32, 632 (1968)
12. Bouma, J. et al.: Soil Sci. Soc. Am. Proc. 32, 362 (1971)
13. Bresler, E.: Water Resourc. Res. 9, 975 (1973)

14. Brown, P. A.: Measurement of water potential with thermocouple psychrometers: construction and application, USDA Forest Service Research Rep., INT-80, 1970
15. Brown, R. H., Konoplyantsev, A. A., Ineson, J., Kovalevsky, V. S. (Eds.): Groundwater studies, UNESCO, Paris 1983 (last supplem.)
16. Cameron, W. M., Pritchard, D. W.: Estuaries, in: The Sea (Ed. Hill, M. N.), Wiley, New York 1963
17. Campbell, G. S.: Soil Sci. 117, 311 (1974)
18. Chorley, R. J.: The hillslope hydrological cycle, in: Hillslope hydrology (Ed. Kirkby, M. J.) 1–42, Wiley, Chichester 1978
19. Chow, V. T.: Open-channel Hydraulics, McGraw-Hill, New York 1959
20. Chow, V. T.: Handbook of Applied Hydrology. McGraw-Hill, New York 1964
21. Colbeck, S. C.: J. Glaciology 11, 369 (1972)
22. Colbeck, S. C.: Advanc. Hydrosci. 11, 165 (1978)
23. Colbeck, S. C.: Water Resourc. Res. 17, 1383 (1981)
24. Collins, E. H.: Trans. Am. Geophysic. Un. p. 1, 624 (1934)
25. Connor, J. J., Brebbia, C. A.: Finite element techniques for fluid flow, Newnes-Butterworths, London 1978
26. Crawford, N. H., Donigian, A. S.: Pesticide Transport and runoff model for agricultural lands, EPA-660/2-74-013, U.S. Environmental Protection Agency, Washington 1973
27. Crawford, N. H., Linsley, R. K.: Digital simulation in hydrology: Stanford Watershed Model IV, Dept. Civ. Eng. Stanford Univ., Techn. Rep. 39 (1966)
28. Dalton, F. N., Rawlins, S. L.: Soil Sci. 105, 12 (1968)
29. Danish Hydraulic Institute: Danish hydraulics 5, 4 (1984)
30. Darcy, H.: Les Fontaines Publiques de la Ville de Dijon, Dalmont, Paris 1856
31. Dawdy, D. R., O'Donnell, T.: Proc. ASCE HY4, 91, 123 (1965)
32. De Haar, U.: Beitr. Hydrol. 2, 85 (1974)
33. De Ploey, J. (Ed.): Rainfall simulation, runoff and soil erosion, Catena Suppl. 4, 1983
34. De Ridder, N. A. (Ed.): Drainage principles and applications, vol. III: Surveys and investigations, Int. Inst. Land Reclamation and Improvement, Wageningen 1974
35. Deutsche Forschungsgemeinschaft (Hrsg.): Theoretische Hydrologie H. 1: Stochastische Verfahren, Boldt, Boppard 1971
36. Deutsche Forschungsgemeinschaft (Hrsg.): Theoretische Hydrologie H. 2: Deterministische Verfahren, Boldt, Boppard 1975
37. De Wiest, R. J. M.: Geohydrology, Wiley, New York 1965
38. De Wit, C. T., van Keulen, H.: Simulation of transport processes in soils, pudoc, Wageningen 1975
39. Donigian, A. S., Imhoff, J. C., Bicknell, B. R., Kittle, J. L.: Application guide for hydrological simulation program – FORTRAN (HSPF), EPA-600/3-84-065, U.S. Environmental Protection Agency, Athens 1984
40. Douglas, J. F.: Solution of Problems in Fluid Mechanics, 2 parts, Pitman, London 1982
41. Dracos, T.: Hydrologie – Eine Einführung für Ingenieure, Springer, Wien 1980
42. Dunne, T., Leopold, L. B.: Water in Environmental Planning, W. H. Freeman & Co., San Francisco 1978
43. Dyck, S.: Angewandte Hydrologie, 2 Bde., Ernst & Sohn, Berlin 1980^2
44. Dyck, S., Peschke, G.: Grundlagen der Hydrologie, Ernst & Sohn, Berlin 1983
45. Eagleson, P. S.: Dynamic Hydrology, McGraw-Hill, New York 1970
46. Elrick, D. E.: Solute transport during infiltration, in: Applications of soil physics (Ed. Hillel, D.), Academic Press, New York 1980
47. Elrick, D. E., Laryea, K. B., Groenevelt, P. H.: Soil Sci. Am. J. 43, 856 (1979)
48. Emmett, W. W.: Overland flow, in: Hillslope hydrology (Ed. Kirkby, M. J.) 145–176, Wiley, Chichester 1978
49. Fiering, M. B., Jackson, B. B.: Synthetic Streamflows, Am. Geophysic. Un., Washington 1971
50. Fleming, G.: Computer Simulation Techniques in Hydrology, Elsevier, Amsterdam 1975
51. Förstner, U., Müller, G.: Schwermetalle in Flüssen und Seen als Ausdruck der Umweltverschmutzung, Springer, Berlin 1974

52. Foster, G. R., Meyer, L. D.: A closed form soil erosion equation for upland areas, in: Sedimentation Symposium to Honour Professor Hans Albert Einstein (Ed. Shen, H. W.) 12.1–12.19, Colorado State Univ. Fort Collins 1972
53. Freeze, R. A.: Mathematical models of hillslope hydrology, in: Hillslope hydrology (Ed. Kirkby, M. J.) 177–225, Wiley, Chichester 1978
54. Fried, J. J.: Groundwater Pollution, Elsevier, Amsterdam 1975
55. Frissel, M. J., Reininger, P.: Simulation of accumulation and leaching in soils, pudoc, Wageningen 1974
56. Gardner, W. R.: Soil Sci. 85, 228 (1958)
57. Giesel, W., Lorch, S., Tenger, M.: Water flow calculation by means of gamma absorption and tensiometer field measurement in the unsaturated soil profile, in: Isotope Hydrology (Ed. IAEA), Int. Atomic Energy Agency, Vienna 1970
58. Graf, H. W.: Hydraulics of Sediment Transport, McGraw-Hill, New York 1971
59. Green, H. W., Ampt, G. A.: J. Agr. Sci. 4, 1 (1911)
60. Gregory, K. J., Walling, D. E.: Drainage Basin Form and Process: A Geomorphological Approach, Arnold, London 1973
61. Groenevelt, P. H., Kijne, J. W.: Physics of soil moisture, in: Drainage principals and applications (Ed. Int. Inst. Land Reclamation), vol. 1, Int. Inst. Land Reclamation and Improvement, Wageningen, 123–151 (1972)
62. Hálek, V., Švec, J.: Groundwater Hydraulics, Elsevier, Amsterdam 1979
63. Hansen, D. V., Rattray, M. J.: Limnol. Oceanogr. 11, 319 (1966)
64. Hartge, K. H.: Die physikalische Untersuchung von Böden, Enke, Stuttgart 1971
65. Hartge, K. H.: Einführung in die Bodenphysik, Enke, Stuttgart 1978
66. HEC (Hydrologic Engineering Center): U.S. Army Corps of Engineers Computer Support Center, Various computer programs for hydrologic, hydraulic, water control systems analysis, Davis, CA, U.S.A.
67. Helliwell, P. R. (Ed.): Urban storm drainage, Pentech Press, London 1978
68. Herrmann, R.: Einführung in die Hydrologie, Teubner, Stuttgart 1977
69. Herrmann, R.: Water, Air and Soil Pollution 16, 445 (1981)
70. Herrmann, R. (Ed.): International Workshop on 'Transport of Organic Micropollutants in the Hydrological Cycle', Fresenius' Z. Anal. Chem. 319 (2), 113 (1984)
71. Herrmann, R., Hübner, D.: Netherlands J. Sea Res. 15, 362 (1982)
72. Herschy, R. W. (Ed.): Hydrometry – Principles and Practices, Wiley, Chichester 1978
73. Hill, E. D., Parlange, J.-Y.: Soil Sci. Soc. Am. Proc. 36, 697 (1972)
74. Hillel, D.: Soil and Water: Physical Principles and Processes, Academic Press, New York 1971
75. Hillel, D.: Applications of soil physics, Academic Press, London 1980
76. Hillel, D.: Introduction to soil physics, Academic Press, Orlando 1982
77. Hillel, D., Gardner, W. R.: Soil Sci. 109, 149 (1970a)
78. Hillel, D., Gardner, W. R.: Soil Sci. 109, 410 (1970b)
79. Hillel, D., Hornberger, G. M.: Soil Sci. Soc. Am. J. 43, 434 (1979)
80. Hillel, D., Talpaz, H.: Soil Sci. 123, 54 (1977)
81. Holtan, H. N.: A concept for infiltration estimates in watershed engineering. U.S. Dept. Agr., Agr. Res. Service Publ., Beltsville, Maryland 1961
82. Hornung, U., Messing, W.: Poröse Medien – Methoden und Simulation, Verl. Beiträge zur Hydrologie, Kirchzarten 1984
83. Horton, R. E.: Trans. Am. Geophysic. Un. 446, 14 (1933)
84. Horton, R. E.: Soil Sci. Soc. Am. Proc. 5, 339 (1940)
85. Horton, R. E.: Bull. Geol. Soc. Am. 3, 275 (1945)
86. Jackson, R. D.: Soil Sci. Soc. Am. Proc. 28, part I 172, part II 464 (1964)
87. Jansen, P. Ph. et al. (Eds.): Principles of river engineering: the non-tidal alluvial river, Pitman, London 1979
88. Johannesen, M., Hendriksen, A.: Water Resourc. Res. 14, 615 (1978)
89. Jordan, P.: Water Resourc. Res. 19, 971 (1983a)
90. Jordan, P.: Water Resourc. Res. 19, 979 (1983b)
91. Keller, R.: Gewässer und Wasserhaushalt des Festlandes, Haude & Spener, Berlin 1961
92. Keller, R. (Hrsg.): Hydrologischer Atlas der Bundesrepublik Deutschland, Boldt-Verlag, Boppard 1979

93. Khan, M. A. Q.: Pesticides in Aquatic Environments, Plenum, New York 1977
94. Kijne, J. W.: Determining evapotranspiration, in: Drainage principals and applications (Ed. Int. Inst. Land Reclamation), vol. 3, Int. Inst. Land Reclamation and Improvement, Wageningen, 53–111 (1974)
95. Kirkby, M. J.: Hillslope hydrology, Wiley, Chichester 1978 a
96. Kirkby, M. J.: Implications for sediment transport, in: Hillslope hydrology (Ed. Kirkby, M. J.), 325–363, Wiley, Chichester 1978 b
97. Kirkby, M. J., Chorley, R. J.: Bull, Intern. Assoc. Sci. Hydrology 12, 5 (1967)
98. Kisiel, C. C., Duckstein, L.: Ground-water models, in: Systems approach to water management (Ed. Biswas, A. K.), McGraw-Hill, New York 1976
99. Klute, A.: Laboratory measurement of hydraulic conductivity of unsaturated soil, in: Methods of soil analysis (Ed. Black, C. A.), part 1, Am Soc. Agronomy, Madison, 253–261 (1965)
100. Konikow, L. F., Bredehoeft, J. D.: Computer model of two-dimensional solute transport and dispersion in groundwater, in: Automated Data Processing and Computations (Ed. U.S. Geolog. Survey), Chapter C2, Washington 1978
101. Korte, F.: Ökologische Chemie, Thieme, Stuttgart 1980
102. Kostiakov, A. N.: On the dynamics of the coefficient of water-percolation in soils and on the necessity of studying it from a dynamic point of view for purposes of amelioration, Trans. Com. Int. Soc. Soil Sci., 6th Moscow A, 12 (1932)
103. Lange, O. L., Kappen, L., Schulze, E.-D. (Eds.): Water and plant life, Springer, Berlin 1976
104. Langguth, H.-R., Voigt, R.: Hydrogeologische Methoden, Springer, Berlin 1980
105. Lazaro, T.: Urban Hydrology, Ann Arbor Science, Ann Arbor 1979
106. Leeder, M. R.: Sedimentology, Process and Product George Allen and Unwin, London 1982
107. Leendertse, J. J., Gritton, E. C.: A water-quality simulation model for well mixed estuaries and coastal seas, 3 vol., Rand Corporation, New York 1970 and 1971
108. Linsley, R. K.: Rainfall-Runoff Models, in: Systems approach to water management (Ed. Biswas, A. K.), McGraw-Hill, New York 1976
109. Loucks, D. P.: Surface-water quantity management models, in: Systems approach to water management (Ed. Biswas, A. K.), McGraw-Hill, New York 1976 a
110. Loucks, D. P.: Surface-water quality management models, in: Systems approach to water management (Ed. Biswas, A. K.), McGraw-Hill, New York 1976 b
111. Matalas, N. C., Wallis, J. R.: Generation of synthetic flow sequences, in: Systems approach to water management (Ed. Biswas, A. K.), McGraw-Hill, New York 1976
112. Mattheß, G.: Die Beschaffenheit des Grundwassers, Borntraeger, Berlin 1973
113. Mattheß, G., Ubell, K.: Allgemeine Hydrogeologie – Grundwasserhaushalt, Borntraeger, Berlin 1983
114. Miller, E. E., Klute, A.: Dynamics of soil water, in: Irrigation of Agricultural Lands (Ed. ASA), 209–244, Am. Soc. Agron. Monogr. 11, Madison 1967
115. Molz, F. J.: Water Resourc. Res. 17, 1245 (1981)
116. Monteith, J. L.: Principles of environmental physics, Arnold, London 1973
117. Monteith, J. L.: Grundzüge der Umweltphysik, Steinkopf, Darmstadt 1978
118. Moser, H., Rauert, W.: Isotopenmethoden in der Hydrologie, Borntraeger, Berlin 1980
119. Mualem, Y.: Water Resourc. Res. 12, 513 (1976)
120. Musgrave, G.: How much of the rain enters the soil? in: Water (Ed. Stefferud, A.), U.S. Dept. Agricult. Yearbook, Washington 1955
121. Nash, J. E.: Proc. Inst. Civ. Eng. 17, 249 (1960)
122. NCASI (Nat. Counc. Paper Industr. Air Stream Improvem.): The mathematical water quality model QUAL-II, Techn. Bull. 391, New York 1982
123. Neff, J. M.: Polycyclic aromatic hydrocarbons in the aquatic environment, Applied Science Pub., Barking 1979
124. Officer, C. B.: Physical Oceanography of Estuaries (and Associated Coastal Waters), Wiley, New York 1976
125. Olausson, E., Cato, I. (Eds.): Chemistry and Biogeochemistry of Estuaries, Wiley, Chichester 1980
126. Orlob, T.: Estuarial Models, in: Systems approach to water management (Ed. Biswas, A. K.), McGraw-Hill, New York 1976
127. Overton, D., Meadows, M.: Stormwater Modeling, Academic Press, New York 1976

128. Owens, W. M.: Trans. Am. Soc. Civil Engrs. 119, 1157 (1954)
129. Parsons, O. A.: Depths of overland flow, Soil Conserv. Ser. Tech. Paper 82, 33pp (1949)
130. Penman, H. L.: Proc. R. Soc. A 194, 120 (1948)
131. Penman, H. L.: J. Soil Sci. 1, 74 (1949)
132. Philip, J. R.: Soil Sci. 84, 257 (1957a)
133. Philip, J. R.: J. Meteorol. 14, 354 (1957b)
134. Philip, J. R.: J. Geophys. Res. 69, 1553 (1964)
135. Philip, J. R.: Theory of infiltration. Adv. Hydrosci. 5, 215 (1969)
136. Philip, J. R.: Soil Sci. Soc. Am. Proc. 39, 1042 (1975)
137. Plate, E. J. et al.: Ablauf von Hochwasserwellen in Gerinnen, Schriftenr. Kurator. Wasser u. Kulturbauwesen H 27, Paul Parey, Hamburg 1980
138. Price, A. J., Dunne, T., Colbeck, S. C.: Energy balance and runoff from a subarctic snowpack, Cold Regions Res. Engin. Lab. Rep. 76–27, Hanover 1976
139. Raats, P. A. C.: Soil Sci. Soc. Am. Proc. 37, 681 (1973)
140. Raudkivi, A. J.: Loose Boundary Hydraulics, Pergamon Press, Oxford 1976
141. Raudkivi, A. J.: Hydrology, Pergamon, Oxford 1979
142. Rawitz, E., Margolin, M., Hillel, D.: Soil Sci. Soc. Am. Proc. 36, 533 (1972)
143. Remson, I., Hornberger, G. M., Molz, F. J.: Numerical methods in subsurface hydrology, Wiley-Interscience, New York 1971
144. Richards, L. A.: Physical condition of water in soil, in: Methods of soil analysis (Ed. Black, C. A.), part 1, Am. Soc. Agron., Madison, 128–152 (1965)
145. Richards, S. J.: Soil suction measurements with tensiometers, in: Methods of soil analysis (Ed. Black, C. A.), part 1, Am. Soc. Agron., Madison, 153–163 (1965)
146. Rinaldi, S., Soncini-Sessa, R., Stehfest, H., Tamura, H.: Modeling and control of river quality, McGraw-Hill, New York 1979
147. Rodi, W.: Turbulence models and their application in hydraulics, State of the art paper, Int. Assoc. Hydraulic Res., Delft 1980
148. Rose, C. W.: Agricultural physics, Pergamon, Oxford 1966
149. Rose, C. W., Byrne, G. F., Begg, J. E.: An accurate hydraulic lysimeter with remote weight recording. CSIRO Div. Land Res. Surv., Tech. Paper 27, Canberra 1966
150. Rovey, E. W., Woolhiser, D. A., Smith, R. E.: A distributed kinematic model of upland watersheds, Colorado State University, Fort Collins, Hydrology Pap. 93, 1977
151. Rubin, J.: Water Resourc. Res. 2, 739 (1966)
152. Salomons, W., Förstner, U.: Metals in the Hydrocycle, Springer, Berlin 1984
153. Sanders, T. et al.: Design of networks for monitoring water quality, Water Resourc. Pub., Littleton 1983
154. Schick, A. P. (Ed.): Channel processes – water, sediment, catchment controls, Catena Suppl. 5, 1984
155. Schlichting, H.: Boundary Layer Theory, McGraw-Hill, New York 1968[6]
156. Schmidt, J.: Mitt. D. Bdkl. Ges. 39, 139 (1984)
157. Schumm, S. A.: Bull. Geol. Soc. Am. 73, 719 (1962)
158. Segol, G.: A three-dimensional Galerkin-finite element model for the analysis of contaminant transport in saturated unsaturated porous media, in: Finite Elements in Water Resources (Eds. Gray, W. G. et al.), Pentech Press, London 1977
159. Shaw, E. M.: Hydrology in Practice, Van Nordstrand Reinhold (UK), Wokingham 1983
160. Shen, H. W., Li, R.-M.: Proc. Am. Soc. Civil Engrs. 99 (HY5), 771 (1973)
161. Shen, H. W., Yen, B. C.: J. Hydrology 68, 333 (1984)
162. Singh, K. P.: Water Resour. Bull. 12, 381 (1976)
163. Slatyer, R. O.: Plant-water relationships, Academic press, London 1967
164. Straub, L. G.: Trans. Am. Geophysic. Un. 20, 649 (1939)
165. Streeter, V. L.: Handbook of fluid dynamics, McGraw-Hill, New York 1961
166. Strelkoff, T.: J. Hydraul. Div. Am. Soc. Civil Engrs. 94 (HY3), 861 (1969)
167. Strelkoff, T.: J. Hydraul. Div. Am. Soc. Civil Engrs. 95 (HY6), 1331 (1970)
168. Stumm, W., Morgan, J. J.: Aquatic Chemistry, Wiley-Interscience, New York 2nd ed. 1981
169. Sündermann, J., Holz, K.-P. (Eds.): Mathematical Modelling of Estuarine Physics. Lecture Notes on Coastal and Estuarine Studies, Springer, Berlin 1980

170. Swartzendruber, D., Hillel, D.: The physics of infiltration, in: Physics of Soil, Water and Salts in Ecosystems (Eds. Hadas, A. et al.), Springer, Berlin and New York 1973
171. Thibodeaux, L. J.: Chemodynamics, Wiley, New York 1979
172. Thomas, H. A., Fiering, M. B.: Mathematical Synthesis of Streamflow Sequences for the Analysis of River Basins by Simulation, in: Design of Water Resources Systems (Eds. Maas, A. et al.), Harvard Univ. Press, Cambridge, Ma. 1962
173. Tinsley, I. J.: Chemical Concepts in Pollutant Behavior, Wiley, New York 1979
174. Todd, D. K.: Groundwater Hydrology, Wiley, New York 1959
175. Topp, G. C.: Soil Sci. Soc. Am. Proc. 33, 645 (1969)
176. Trescott, P. C.: Documentation of finite-difference model for simulation of three-dimensional ground-water flow, U.S. Geol. Survey Open File Rep. 75–438, Reston VA 1976
177. Trescott, P. C., Pinder, G. F., Larson, S. P.: Finite-difference model for aquifer simulation in two dimensions with results of numerical experiments, in: Automated Data Processing and Computations (Ed. U.S. Geol. Survey), Chapter C1, Washington 1976
178. Tritton, D. J.: Physical fluid dynamics, Van Norstrand Reinhold, London 1977
179. Uhlmann, D.: Hydrobiologie, G. Fischer, Stuttgart 1982[2]
180. Ungs, M., Cleary, R. W., Boersma, L., Yingjajaval, S.: The description of transfer of water and chemicals through soils, in: Land as a Wast Management Alternative (Ed. Loehr, R. C.), Ann Arbor Science, Ann Arbor 1976
181. Van Genuchten, M. T.: Soil Sci. Soc. Am. J. 44, 892 (1980)
182. Van Genuchten, M. T., Pinder, G. F., Sankin, W. F.: Modelling of leachate and soil interactions in an aquifer, in: Management of Gas and Leachate in Landfills (Ed. Banerji, S. K.), EPA-600/9-77-026, U.S. Environmental Protection Agency, Cincinatti 1977
183. Verruijt, A.: Theory of groundwater flow, Macmillan, London 1970
184. Vischer, D., Huber, A.: Wasserbau, Springer, Berlin 1979
185. Wankiewicz, A.: A review of water movement in snow, in: Proceedings, Modeling of Snow Cover Runoff (Eds. Colbeck, S. C., Ray, M.), Cold Regions Res. Engin. Lab., Hanover, 222 (1979)
186. Warrick, A. W., Nielsen, D. R.: Spatial Variability of Soil Physical Properties in the Field, in: Application of soil physics (Hillel, D.), Academic Press, London 1980
187. Weyman, D. R.: Bull. Internat. Assoc. Sci. Hydrology 15, 25 (1970)
188. Wiesner, C. J.: Hydrometeorology, Chapman and Hall, London 1970
189. Wilhelm, F.: Schnee und Gletscherkunde, De Gruyter, Berlin 1975
190. Williams, J. R.: Water Resourc. Res. 14, 659 (1978)
191. Wischmeier, W. H., Smith, D. D.: A universal soil loss equation to guide conservation farm planning, Trans. Seventh Intern. Congress Soil Sci. 1, 418 (1960)
192. WMO (World Meteorological Organization): Manual for Depth – Area – Duration Analysis of Storm Precipitation, Geneva, WMO No. 237, TP 129 (1969)
193. Woo, D.-C., Brater, E. F.: Proc. Am. Soc. Civil Engrs. 88 (HY6), 31 (1962)
194. Yevjevich, V.: Probabilities and statistics in hydrology, Water Res. Publ., Fort Collins 1972
195. Yoon, Y. N., Wenzel, H. G.: Proc. Am. Soc. Civil Engrs. 97 (HY9), 1367 (1971)
196. Youngs, E. G.: Soil Sci. 109, 307 (1964)
197. Zuzel, J. F., Cox, L. M.: Water Resourc. Res. 11, 174 (1975)

Outdoor Ponds: Their Construction, Management, and Use in Experimental Ecotoxicology

N. O. Crossland

Shell Research Limited, Sittingbourne Research Centre
Sittingbourne, Kent ME9 8AG, United Kingdom

C. J. M. Wolff

Shell Research B.V., Koninklijke/Shell-Laboratorium
Postbus 3003, NL-1003 AA Amsterdam, The Netherlands

Summary

Experiments in outdoor ponds can play a useful role in ecotoxicology, intermediate between laboratory testing and environmental monitoring. Experimental ponds should closely simulate the natural environment and variation between them should be kept to a minimum. Methods of constructing and managing outdoor ponds so as to achieve these aims are discussed.

The use of outdoor ponds to evaluate methods of risk assessment is illustrated by reference to controlled field experiments, where fate and biological effects of chemicals in ponds are compared with predictions based on process analyses, mathematical models and the results of single-species toxicity tests.

Introduction

Outdoor experimental ponds have been used by ecologists to study the effects of nutrient enrichment on freshwater ecosystems [1–3], the productivity of sewage-fertilised fish ponds [4] and predator-prey interactions in freshwater communities [1, 5]. In addition to fundamental work of this nature, outdoor ponds have been widely used to study the dispersion, persistence, bioaccumulation, and toxic effects of oil and chemicals in natural or semi-natural environments [5–31]. Most of these studies have been concerned with pesticides, where the objective has been to obtain information on the fate and effects of chemicals under conditions that closely simulate the natural environment. Recently, there has been increasing interest in the use of outdoor ponds and streams to evaluate methods for predicting environmental fate and biological effects from data generated in the laboratory [32–34]. In this chapter we shall be concerned with the construction, management, and use of outdoor ponds for this purpose.

In an earlier volume in this series the use of mathematical models to predict the behaviour of chemicals in aquatic environments was described [35]. The need to predict environmental exposure with the aid of such models has been a stimulus to their development and significant progress has been made in recent years [36–39]. However, there have so far been relatively few attempts to validate the models by comparing their predictions with results obtained in laboratory microcosms or in field situations. Pritchard [40] has reviewed the use of microcosms for this purpose. He emphasised that results or data produced from a microcosm study are only as good as the background information required to interpret the results. The rate of processes such as volatilisation, hydrolysis, phototransformation, and biodegradation can be markedly affected by the environmental conditions of the microcosm. Therefore, when designing experiments to evaluate the fate of chemicals it is essential to have some prior knowledge of the processes that may affect transport and degradation of a particular chemical. Such knowledge may then be used as a basis for controlling or monitoring environmental variables that might affect the rates of these processes. The same basic approach is required when designing experiments in outdoor ponds. The value of a particular experiment can only be as good as the background information required to interpret the results. Since outdoor environmental conditions cannot be controlled it is essential that there is adequate provision for monitoring relevant environmental variables.

Methods for predicting biological effects of chemicals in natural freshwater environments rely heavily on the results of single-species toxicity tests. However, in only a few instances have scientifically valid comparisons been made between the results of single-species toxicity tests and biological effects observed in natural environments. A wide range of species exists in all natural environments and only a few of these, at best, will have been used in single-species toxicity tests for a particular chemical. The susceptibility of some of the species present in natural environments may differ quite markedly from that of the species used in toxicity tests. Furthermore, secondary biological effects can occur in natural environments [41] and it is impossible to predict these on the basis of single-species toxicity tests. It is clear, therefore, that data generated from laboratory tests may need to be complemented by field tests, at least for those chemicals that might present toxic hazards at, or near to, concentrations that could occur in the environment.

Construction and Management of Outdoor Ponds

Outdoor ponds of various shapes and sizes have been used in environmental toxicology. In general, ponds with a small volume, less than 5 m³, have not proved to be very satisfactory for most purposes. Small artificial ponds are inherently unstable ecosystems that are subject to rapid successional changes in their flora and fauna. These changes may lead to wide variation in the species composition of a series of experimental units thus rendering them unsuitable for experimental purposes. Such was the experience of Caspers and Hamburger [42] who found that variation between 2 m^3 artificial ponds was so great that they were unable to detect the effects of relatively toxic wastewater against a background of high biological variation.

At the upper end of the size range Hall et al. [1] have described a system of 50 experimental ponds built at Cornell University, Ithaca, New York. The volume of each pond was 650 m^3, the maximum depth was 1.3 m and the surface area 700 m^2. Ponds of this size are typical of those used for rearing fish and they closely simulate natural freshwater environments in agricultural areas. Replicated experiments carried out using such ponds have yielded basic insights into the structure and function of freshwater ecosystems and they provide a valuable model for the design of outdoor pond experiments in aquatic toxicology. Outdoor ponds used at the U.S. Fish and Wildlife Laboratory, Columbia, Missouri were built to similar specifications and have been used to assess the fate and effects of various herbicides and insecticides on freshwater communities under realistic field conditions [26, 43, 44].

Outdoor ponds intermediate in size between these two extremes are probably optimal for use as a basic field plot in aquatic toxicology. The smaller the basic unit, the lower the capital and operational costs. On the other hand, the larger the basic unit, the more stable it is likely to be and the more closely it will simulate the natural environment. Ponds with a volume of about 10 to 50 m^3 are big enough to provide a habitat for a wide variety of organisms, including small fish populations, although if fish are important it may be unwise to opt for the lower end of this suggested size range.

We have constructed and developed a system of 40 m^3 ponds that contain most, if not all, of the functional components that are present in larger natural ecosystems. A system of 12 ponds was constructed such that, when not in use for experimental purposes, they were interconnected by pipes. The water between ponds was occasionally mixed by circulating it with the aid of a pump, thus promoting the uniform distribution of organisms among the ponds. Each pond is 5 m wide, 10 m long, and has a maximum depth of 1 m. The longer, dividing walls are vertical and made from concrete. The shorter sides of the ponds slope at an angle of about 45° and they consist, as do the bottom of the ponds, of a mixture of clay, alluvial silt and organic debris. Clay, removed during excavation, was used to replace more porous substrates such as gravel intrusions, thus forming a relatively impermeable bottom layer.

We have adopted various management strategies in order to develop and maintain a system of similar ponds that will support a variety of aquatic plants, invertebrates and fish. Since they are situated in the floodplain of an unpolluted and productive chalk stream we considered it unnecessary to colonise the ponds artificially by seeding them with plants or animals. Our approach was to create a suitable habitat and to rely on natural colonisation of the ponds by organisms from the nearby stream and elsewhere. After filling the newly-dug ponds with stream water we added a supply of organic matter and nutrients in the form of partially-rotted bales of hay. Within a few weeks large numbers of zooplankters were present and within a few months a wide variety of aquatic insects were seen. The Diptera were early colonisers, particularly Chaoboridae and Chironomidae. Within a year 19 families of macroinvertebrates were present although, at this stage, the diversity of the macroinvertebrate community was less than that in neighbouring, more mature, ponds.

During the second year of their development the diversity of the plant and animal communities increased. The dominance of planktonic, unicellular algae was succeeded by the development of filamentous algae and vascular plants. Development of this macrophytic vegetation was associated with the colonisation and development of large populations of mayfly larvae and other aquatic insects that are, to a greater or lesser extent, dependent on these plants for food or shelter. By the end of the second year the pond communities were sufficiently diverse and productive for our purposes, i.e. they contained species that are representative of most, if not all, of the important groups of macroinvertebrates that occur in larger, natural, freshwater ecosystems.

Subsequent management has mainly involved various methods of control of the aquatic vegetation. Filamentous algae can grow very quickly and can destabilise the natural balance of pond ecosystems. The vascular plants grow more slowly and they can have a stabilising effect because they remove excess nutrients from the water during their growing season and thus they may inhibit the development of undesirable algal blooms. However, the vascular plants may also need to be controlled, otherwise dense growths in some ponds, but not in others, can lead to unacceptably high variation between ponds. At various times, depending on its species composition, its density and seasonal considerations, we have attempted to control the aquatic vegetation by using herbicides, by introducing grass carp (*Ctenopharyngodon idella*) and by cutting and removing it by hand.

The use of the bipyridylium herbicides, diquat and paraquat, for control of submerged aquatic vegetation is well established. They are contact herbicides that destroy green plant tissue with relatively little translocation in the plant. They are lost from the water within a few days after treatment and become inactivated by sorption onto sediments [45]. Their lack of persistence in water and plants and lack of toxicity to invertebrates and fish at herbicidal dosages [45–47] make them suitable for vegetation control in experimental ponds where other control methods are ineffective. We found that paraquat was useful in controlling dense growths of submerged vegetation that could not easily be controlled in any other way.

Control of marginal rushes and reeds such as *Juncus, Typha,* and *Sparganium* can be achieved through spot applications of the herbicides dalapon or glyphosate. These herbicides are translocated from foliage and roots and are non-toxic to fish and invertebrates at dosages used to control the vegetation. We have used dalapon when carrying out spot treatments of rushes during the autumn, when the ponds were not being used for experimental purposes. This treatment was followed by mechanical cutting of the remaining aerial parts of the rushes in spring. Clearly, the use of herbicides in experimental facilities of this nature is undesirable and we have minimised their use as far as possible. Whenever they were necessary, herbicide treatments were carried out at least six months before using the ponds for experiments.

The use of grass carp might seem an attractive alternative to the use of herbicides for control of aquatic vegetation. However, it is difficult to control the biomass of these herbivorous fish in relation to the biomass and turnover rate of their food supply. Their effect on the submerged vegetation of individual ponds is therefore unpredictable and they are not a sufficiently reliable tool for vegetation management in small, experimental ponds. Attempts to use them for this purpose may lead to an increase in the variability of vegetation between ponds, especially since grass carp do not feed equally on all species of aquatic plants.

Hand-cutting and removal of the submerged vegetation is laborious and is an ineffective method of controlling certain species, particularly filamentous algae. Nevertheless, we have found it to be a useful technique for controlling the density of dominant species of vascular plants, particularly in the spring before the start of experimental work. By this means it is possible to ensure that similar amounts of aquatic macrophytes are present in the ponds at a time when invertebrates are actively growing and reproducing. This method may thus be used to minimise the variability of habitat between ponds.

Design of Pond Experiments

The design of any experiment is dependent on its specific objectives. In ecotoxicology the objectives of pond experiments may vary depending, among other things, on the end use of a particular chemical, its physicochemical characteristics and toxicological properties. Field studies with a particular chemical would not normally be undertaken until sufficient laboratory work had been carried out to characterise its physico-chemical properties and its toxicity to representative species of aquatic organisms, usually a species of unicellular alga, an invertebrate (e.g. *Daphnia magna*) and a species of fish. On the basis of such laboratory-derived data it should be possible to make a preliminary assessment of potential environmental hazards and in the light of this assessment to decide whether or not to undertake field studies. The role of field-testing in sequential schemes of hazard assessment has been fully discussed elsewhere [48] and will not be elaborated here. It may be noted, however, that field tests will not normally be required for registration of new chemicals unless laboratory data indicate that adverse effects may result from their intended use.

Available laboratory data will usually indicate which process or processes are likely to be involved in the environmental distribution and loss of a particular chemical. The objective of a pond experiment may then be to obtain information on the relative importance of these key loss processes under a particular set of environmental conditions that may affect them. Environmental variables that can affect transport processes, e.g. windspeed, depth of water, depth and organic matter content of sediment, may not vary significantly between ponds, especially if the ponds have been managed in the ways suggested in the previous section. Replication of ponds may therefore be unnecessary in order to study the influence of such variables on transport processes. On the other hand, some of the environmental variables that affect degradation processes, e.g. pH, light absorption by the water, bacterial numbers, can vary considerably between ponds and therefore some replication of ponds is indicated to study the influence of such variables on degradation.

Replication of ponds is essential to evaluate biological effects of chemicals on ecosystems. Careful management prior to treatment can do much to reduce the variability between ponds but, even so, differences of a biological nature will certainly exist. As Hurlbert [41] has pointed out, lack of system replication cannot be offset by taking larger numbers of replicate samples within a system. More intense sampling of this sort may be useful in obtaining more accurate measurement of the differences between systems but it in no way assists a statistical demonstration that observed differences between systems are attributable to chemical treatment. The number of pond replicates required for a particular experiment will depend on the nature of the biological responses to be measured and the variability of the biological parameters of interest between ponds. Thus some pre-treatment data should be obtained in order to assess biological variability between ponds as a guide to experimental design. The greater this variability the more replicates will be required in order to demonstrate statistically significant differences between controls and treatments. It must always be borne in mind that the pond, not the organism, is the experimental unit that is being treated and that the sensitivity of the measurement of the biological responses is dependent on the biological variability between, not within, ponds.

Process Analysis Applied to Outdoor Experiments

As a first step in the design of a pond experiment, a process analysis should be carried out to evaluate the relative importance of various fate processes and to estimate their rate constants. Based on this analysis it should then be possible to assess:
a) which compound-related variables are important and whether available data for these are adequate
b) which environment-related variables need to be monitored
c) in which phases (air, water, sediment, biomass) the compound will be present, and
d) how frequently the concentration of compound in each phase needs to be measured.

In fate modelling practice an environmental system generally is subdivided into a series of compartments, in number varying from 2 to more than 100 [49–52]. Each of these compartments is considered to be homogeneously mixed. Obviously this assumption sets limits to the dimensions of the various compartments depending on the mixing characteristics of the system, the time and geographical scale of the problem and the desired accuracy of the model predictions. For modelling outdoor ponds subdivision into three compartments, one water, one sediment, and one biomass compartment will generally be acceptable (Fig. 1). It is advisable, however, to check regularly the homogeneity within the various compartments and if necessary to further subdivide the compartments.

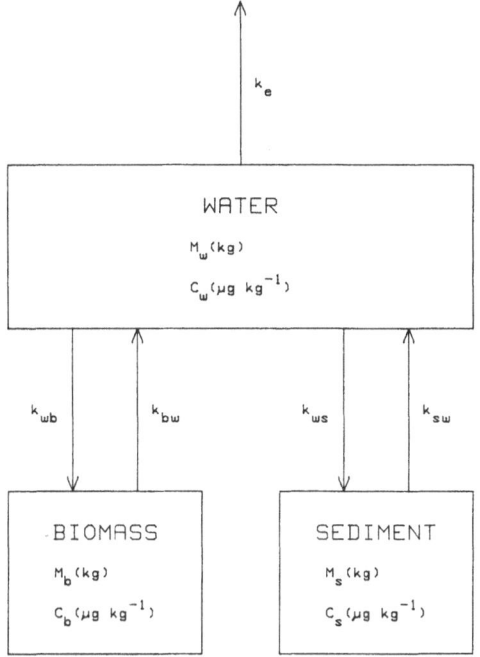

Fig. 1. Three-compartment model for transport of chemicals in outdoor ponds.
M_w, M_b, M_s are mass of water, biota, and sediment, respectively
C_w, C_b, C_s are chemical concentrations in water, biomass, and sediment
k_e is the rate constant for evaporation
k_{wb}, k_{bw}, k_{ws}, k_{sw} are rate constants for transport between water, biomass and sediment

Two types of fate process can be distinguished, viz. transport and transformation. Through compartmentalisation of an environmental system and the assumption of homogeneity within compartments, simplified kinetic descriptions for the processes of transport can be deduced. Kinetic descriptions developed for various transformation processes [36] are basically identical. General kinetic descriptions for the different types of process are presented in Table 1.

Transport Processes

For outdoor ponds only transport from water to air (evaporation), between water and biomass and between water and sediment is expected to be important. The kinetics of these processes are dominated by the dispersion coefficient, the dimensions of the compartments and the partition coefficient of the compound between the two phases [51, 52].

Models to assess the rate of evaporation of dissolved compounds from surface waters have been published [37, 63]. Compound-related parameters dominating this process are Henry's Law constant (H), the diffusion coefficients of the compound in air and water (D_g and D_l) and the dissociation constant (pK_a). Impor-

Table 1. General kinetic equations for transport and transformation processes

Type of process	General kinetic description
Transport within compartments	$\dfrac{dc_1}{dt} = -k_1 (c_1 - c_2)$
Transport between compartments	$\dfrac{dc_1}{dt} = -k_2 (c_1 - Kc_2)$
Transformation	$\dfrac{dc_1}{dt} = -k_3 c_1 [R]$

$\dfrac{dc}{dt}$ is the change of the concentration of the compound with time (mol m^{-3} s^{-1})

$c_{1,2}$ is the concentration of the compound in compartment 1 or 2 (mol m^{-3})
t   is the time (s)
$k_{1,2}$ is the first-order rate constant for transport (s^{-1})
k_3 is the second-order rate constant for transformation (m^3 mol^{-1} s^{-1})
K   is the partition coefficient of the compound between different phases (dimensionless)
[R]   is the concentration of reactive agent responsible for the transformation reaction (mol m^{-3})

tant environment-related parameters are the wind speed, the depth of the pond and the water temperature. The wind speed and profile above small ponds may be strongly influenced by irregularities of the surroundings, such as trees, walls, and banks. It is advisable, therefore, to measure the wind speed close to the water surface.

Transport between water and sediment occurs via the interstitial water phase. The rate of this process is not first-order but decreases with increasing sediment depth. Theoretically, benthic sediment has an unlimited depth. However, by defining the thickness of a layer of "active" sediment an average pseudo-first-order rate constant for transport between water and that layer can be defined, and can be used for modelling. The layer of "active" sediment refers to the upper layer of sediment in which chemicals may be found in measurable quantities. The concentration of chemicals in this layer will decrease with increasing depth [54, 55]. The use of an average concentration for the whole layer of "active" sediment will suffice in most cases but sometimes it may be advisable to further subdivide the layer into different compartments. However, subdivision is often limited by the ability to sample thin slices of sediment. We have found that the thickness of the "active" sediment layer may be taken to be 0.02 m for experiments of 2 to 3 months duration. During longer periods of time (1 to 2 years) compounds may penetrate much deeper [55].

The first-order rate constant for transport between water and sediment, k_2, is determined by the dispersion coefficient and the dimensions of the compartments and is generally independent of compound-related parameters. The value of k_2 for a specific system can be established by the use of a reference compound. The partition coefficient of a compound between water and sediment can be determined in the laboratory. This parameter is the ratio between the concentration of the compound in sediment and that in water at equilibrium. Some empirical relationships between the sediment/water partition coefficient and the octanol/water partition coefficient have been derived [56–58] but doubt exists about their common applicability [59–61].

The mass of the biota in a pond ecosystem is generally small compared to that of water and sediment and therefore the overall fate of a compound is unlikely to be significantly affected by transport between the biota and other compartments. Nevertheless, transport between the water compartment and the biota, which may be divided into several compartments in order to study accumulation via food chains, is of considerable interest in the context of accumulation of persistent chemicals. This subject has been fully discussed elsewhere in the literature [56, 81–89] and will not be elaborated here.

As an example of the use of a reference compound to determine rate constants for the transport processes, we refer to one of our own (unpublished) experiments with the compound 2,5,4'-trichlorobiphenyl (3-CB). From a review of the literature [62] it was evident that the fate of 3-CB in ponds would be determined predominantly by its rate of transport between water, air, sediment, and biota. Measurable degradation of 3-CB was not expected to occur during the period of the experiment (28 days). Three ponds were each treated by subsurface application of a mixture of 3-CB in acetone and water to give a nominal concentration for 3-CB of $14 \ \mu g \ l^{-1}$ in the pond water. Samples of water, sediment, and vegetation were removed from the ponds at various intervals after treatment and analysed for residues of 3-CB. The results of these analyses are given in Table 2.

The data were fitted to a compartmental model in order to investigate the kinetics of transport. Values for the rate constants (k_e, evaporation; k_{ws}, water-sediment; k_{wv}, water-vegetation) were calculated using mass balance equations. In a first series of calculations it was found that the observed concentration-time profile of dissolved 3-CB could not be fitted to the model using a single value of k_e, the rate constant for evaporation, and this variation could not be explained by variations in the wind speed. The experimental period of 28 days was then divided into three shorter periods, viz. 0 to 4 days, 4 to 14 days, and 14 to 28 days, and

Table 2. Concentrations of 3-CB in water ($\mu g \ l^{-1}$), sediment ($\mu g \ g^{-1}$) and vegetation ($\mu g \ g^{-1}$) taken from three outdoor ponds

Time (days)	Mean concentration of 3-CB (and S.D.)		
	Water ($\mu g \ l^{-1}$)	Sediment ($\mu g \ g^{-1}$)	Vegetation ($\mu g \ g^{-1}$)
0	≤ 0.1	≤ 2	≤ 2
0.2 (4 h)	12.2 (1.0)		
1	8.1 (0.3)	≤ 2	
4	3.0 (0.05)	13.7 (8.0)	4.2 (0.72)
7	1.8 (0.05)		
14	0.8 (0.05)	27.2 (10.3)	1.0 (0.32)
28	0.2 (0.05)	39.2 (21.5)	0.24 (0.07)
365		38.0 (23.8)	

Table 3. Calculated values of first-order rate constants in pond experiment with 3-CB

Period (days)	k_e (evaporation) day^{-1}	k_{wv} (water-vegetation) day^{-1}	k_{ws} (water-sediment) day^{-1}
0–4	0.37	0.33	0.014
4–14	0.19	0.32	0.019
14–28	0.10	0.32	0.050

rate constants were calculated separately for each of these periods. The results of these calculations are given in Table 3.

Apparent changes in the value of k_e suggested that enhancement of evaporation probably occurred. This enhancement was attributed to flotation and subsequent evaporation of 3-CB particles, formed when the compound was added to the ponds at a concentration near to the limit of its water solubility. Such a process would strongly influence the rate of loss of 3-CB during the early stages of the experiment. In this case the calculated value of 0.10 for k_e during the period 14–28 days represents the true value for the rate of evaporation of dissolved 3-CB. This value is in good agreement with values of 0.06–0.09 that were predicted from use of the model VRDAMP [37].

It is clear that evaporation was the main process involved in loss of 3-CB from the pond water. It was calculated that this process accounted for 60% loss after four days and 90% loss after 28 days. Sorption of 3-CB onto sediment accounted for 3% loss after four days and 8% loss after 28 days. Mass balance calculations indicated that desorption from vegetation was relatively rapid but desorption from sediment was very slow.

Transformation Processes

The more important transformation processes that may occur in outdoor ponds are hydrolysis (in water and sediment), direct and indirect phototransformation (in water) and aerobic biodegradation (in water and sediment). The kinetics of these processes are determined by the second order rate constant, k_3, a compound-related parameter, and by the concentration of reactive agent, [R], an environment-related parameter (see Table 1). Assuming that no changes occur in the environmental system through the reaction, [R] can be considered constant and the reaction becomes pseudo first-order with the rate constant being $k_3 \times$ [R]. Reaction mechanisms and rates can differ greatly for different forms of a compound. Therefore, for each degradation process pK_a and pH are, respectively, important compound-related and environment-related parameters.

Three types of hydrolysis can be distinguished [36, 56, 63, 64] viz. neutral, specific-acid-catalysed and specific-base-catalysed hydrolysis, and guidelines for measuring rate constants for these processes have been developed [63]. It has been found that hydrolysis can be catalysed by metal ions. These metal ions are normally present in natural sediments but at such low concentrations that they are not expected to play a significant role in most hydrolysis reactions that occur in the environment [64]. For the 1-octyl ester of 2,4-dichlorophenoxyacetic acid, indications were found that the alkaline hydrolysis rate constant decreased by a factor equal to the fraction of the ester associated with humic substances [65].

The rate of direct phototransformation depends on the rate of solar light absorption, i.e. the amount of photons absorbed by the compound per unit of time, and on the quantum yield, i.e. the fraction of the compound that is converted upon absorption of a photon [38, 66, 67]. The rate of solar light absorption can be calculated from the light absorption spectrum of the compound and the number of available photons in each spectral band. So, for this process, the quantum yield and the light absorption spectrum are compound-related parameters. The number of available photons in each spectral band is an environment-related parameter that is determined by a variety of environmental factors [67]. A mathematical model has been developed [38] that can be used to calculate rates of direct phototransformation taking into account the most important of these factors. Results, however, have to be corrected for cloud cover using empirical relationships [68].

Outdoor ponds were used [69, 70] to validate model predictions of the rates of direct phototransformation of pentachlorophenol (PCP) and 3,4-dichloroaniline (DCA). Half-lives of PCP in pond water were predicted to be within the range 1.5 to 3.0 days, in good agreement with observed half-lives of 2.0 to 4.0 days. Most of the variation in half-lives was attributable to variable light attenuation by different pond waters. Half-lives of DCA in pond water were predicted to be within the range 3.5 to 5.8 days, in good agreement with observed half-lives of 4.1 to 6.3 days. The concentration-time profile for DCA in three treated ponds is shown in Fig. 2 to illustrate the effect of variable light attenuation by pond waters on the rate of direct phototransformation.

Indirect phototransformation is the general expression for processes where the compound is transformed through reaction with photochemically-formed reactive species. The occurrence of a variety of these species has been demonstrated in the recent past such as singlet oxygen [71–73], hydrogen peroxide [74], ozone [75], superoxide [74], OH radicals [76], and RO_2 radicals [76]. For modelling purposes the concentration and reactivity towards the compound for all of these reactive species in water should be measured to enable calculation of the pseudo-first-order rate constant ($k_3 \times [R]$). However, this would probably be a waste of effort since the individual pseudo-first-order rate constants are generally very small. Therefore, a test method for measuring the combined effect of all species is advisable, although in this case a series of experiments would need to be performed to establish the influence of environment-related parameters, i.e. the concentration of sensitisers [77].

The second-order rate equation (Table 1) has been demonstrated to be applicable to aerobic biodegradation for a variety of natural waters, for different compounds and for different concentrations of compounds and bacteria [39]. However, the number of bacteria associated with solid surfaces (plants, sediments, etc.) is generally much higher (by a factor of 10^2–10^4) than the number of bacteria in the surrounding water phase. For model predictions these should, therefore, be taken into account [54, 78].

The importance of surface-attached bacteria in degrading MEP in outdoor ponds was demonstrated by a series of experiments carried out in laboratory

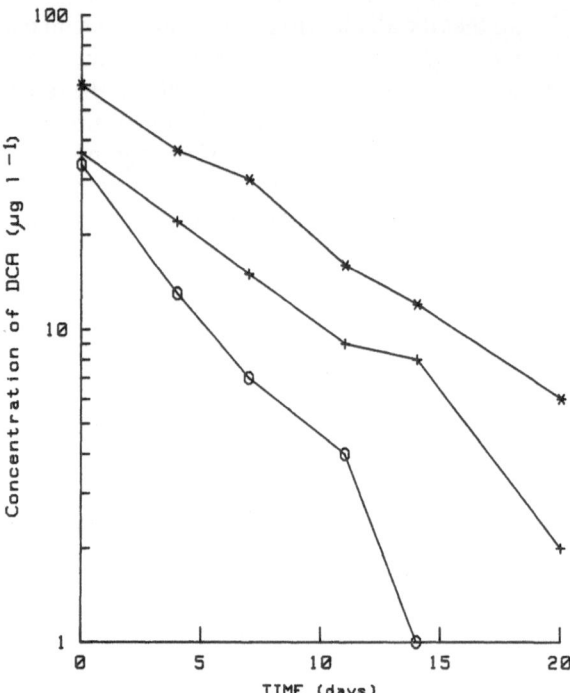

Fig. 2. Concentration-time profile for DCA in three treated ponds

aquaria and in outdoor ponds [78]. The rate of loss of MEP from the water in outdoor ponds was much faster than predictions based on sediment: water partitioning and biodegradation by bacteria suspended in the pond water (Fig. 3). In the light of these results it was decided to investigate the influence of sediments and plants on the rate of loss of MEP in laboratory aquaria. Experiments were then carried out in which MEP was added to the subsurface water of aquaria containing tap water, pond water, tap water plus plants, tap water plus sediment, and tap water plus sediment and plants. The results of these experiments are shown in Fig. 4. The rate of loss of MEP from pond water alone was similar to the expected rate of loss. However, addition of plants or sediments to the aquaria was associated with a much faster rate of loss. It was concluded that biodegradation was the key loss process and that this process was mainly associated with sediment bacteria or bacterial populations attached to plants.

The fact that the observed results in outdoor ponds differed from expectations was very useful in that it led to further laboratory experiments that enabled us to gain useful insights into the important role of bacteria attached to sediments and plants in biodegrading chemicals in aquatic systems. The kinetics of sorptive phenomena were obscured by the relatively rapid biodegradation of sorbed MEP in our experimental systems but absence of MEP in sediment indicated that the rate of biodegradation exceeded the rate of transport to sediments. Thus the latter process may have been important in determining the overall rate of loss of MEP from pond water. We then investigated this further by re-calculating the half-life

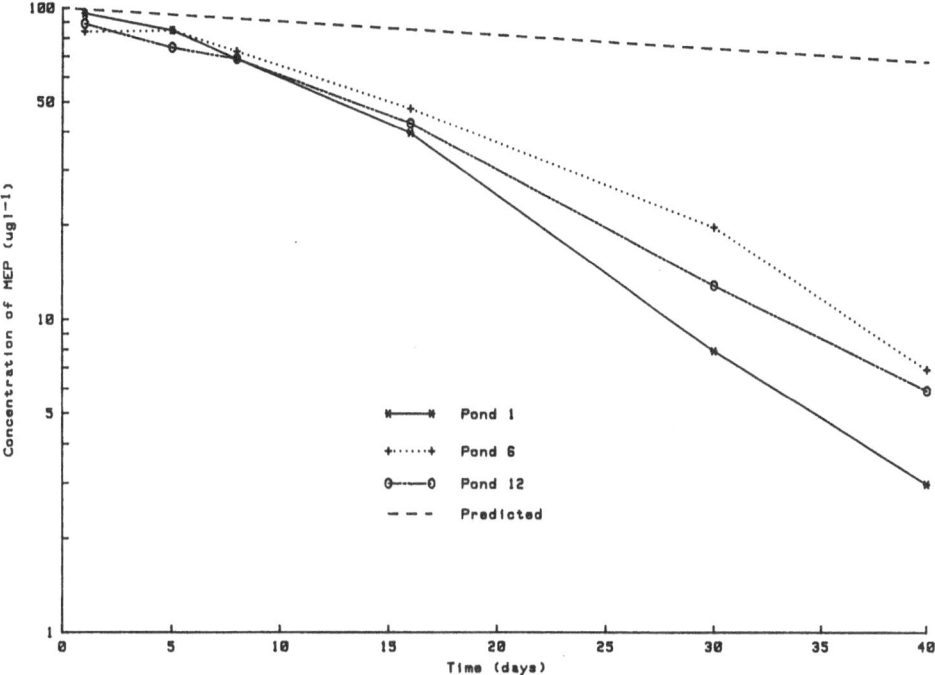

Fig. 3. Loss of methyl parathion from the water of three ponds compared with predictions

of MEP in the outdoor ponds using the compartmental model and rate constants (Table 3) obtained in the experiment, previously described, with 3-CB. We thus obtained calculated half-lives for MEP of 12–24 days, in good agreement with observed half-lives of 9–14 days (Fig. 3).

Biological Effects in Outdoor Pond Experiments

Toxic chemicals can have both direct and indirect effects on aquatic ecosystems [41]. These effects can occur at various levels, viz. at the organism, population, community or ecosystem level. It is possible to investigate effects at all of these levels in outdoor pond experiments and thus a wide variety of biological sampling programmes can be devised. However, unless resources are practically unlimited, it is necessary to choose which organisms and which ecological processes to study with respect to a given chemical. Generally, the choice must be based on relatively limited laboratory data and careful definition of experimental objectives.

Before proceeding with a pond experiment toxicity data for a given chemical should be obtained, either from the literature or from laboratory toxicity tests. These data should include, as a basic minimum, acute LC_{50} values for a species of alga, an invertebrate, and a fish. On the basis of these data it should be possible to decide which species, among those present in the pond community, are likely to be most susceptible to the chemical. Sampling programmes may then be orientated towards the study of effects on populations of the most susceptible species.

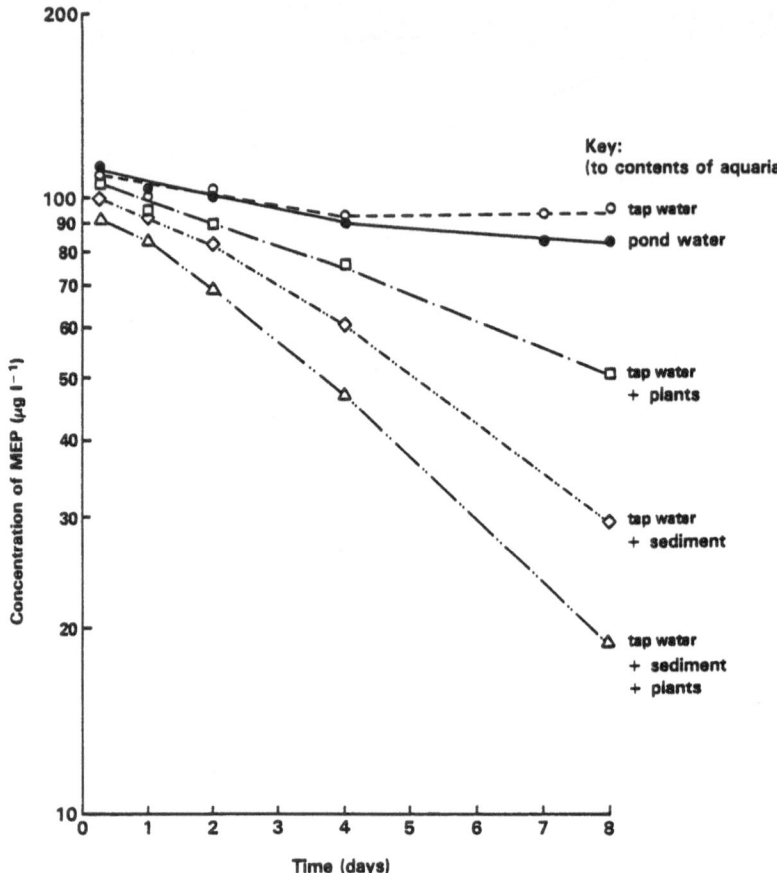

Fig. 4. Loss of methyl parathion from the water in laboratory aquaria

Indirect, or secondary, effects can be more important to the integrity and stability of freshwater ecosystems than direct effects of toxic chemicals. In outdoor pond experiments it has been possible to establish cause-and-effect relationships between chemical treatments and indirect effects such as algal blooms [5, 24, 79], effects on dissolved oxygen concentrations [41] and predator-prey interactions [41]. Such effects are the result of interactions between chemicals, organisms, and their environment. Investigation of indirect effects therefore requires more replication of systems (ponds) than does investigation of direct toxicity.

To illustrate some of these principles we refer to the pond experiment with MEP [78] mentioned in the previous Section. MEP was applied to the subsurface water of three ponds at a target concentration of 100 µg l^{-1} and the resulting concentration-time profile for MEP in each pond is shown in Fig. 3. Samples of zooplankton and macroinvertebrates were taken from the ponds before treatment and at various intervals thereafter until affected populations had recovered. Toxicity of MEP to zooplankton and aquatic insects was similar to predictions based on laboratory toxicity tests and a literature survey. *Daphnia* populations were virtually eliminated (Fig. 5), the Cyclopidae were affected to a much lesser extent,

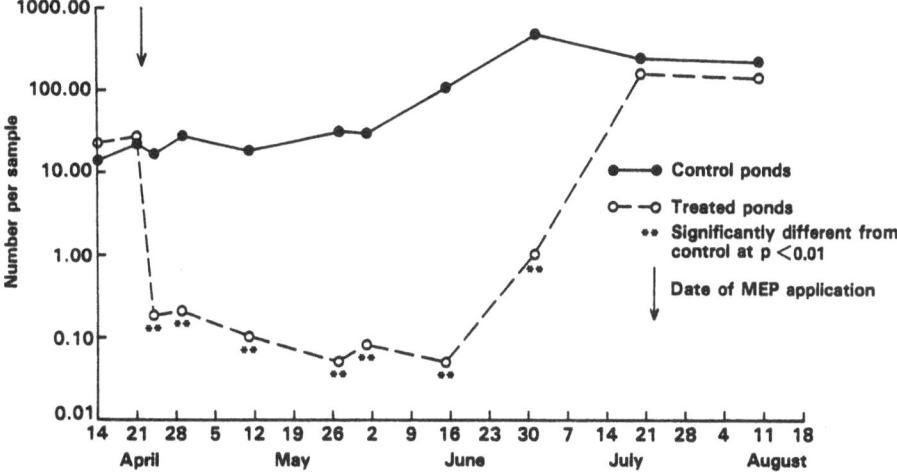

Fig. 5. Effect of MEP on *Daphnia* populations

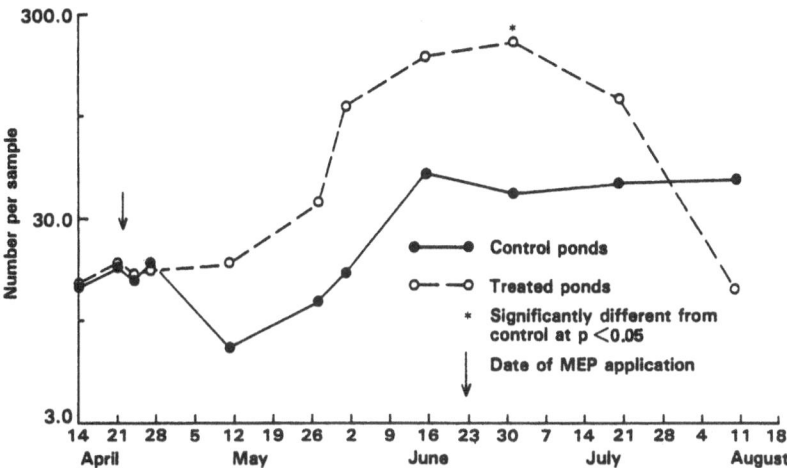

Fig. 6. The rise and fall of diaptomid populations in MEP-treated ponds

while the numbers of Diaptomidae (Fig. 6) were not directly affected. Numbers of free-swimming dipterous and mayfly larvae (Ephemeroptera) were significantly reduced $P < 0.05$) compared with those in control ponds, as were the benthic chironomid larvae.

Recovery of zooplankton and aquatic insect populations occurred very soon after concentrations of MEP had declined below threshold toxicity levels (0.01–1 $\mu g\, l^{-1}$ for the most susceptible species). On the 1st July, 70 days after treatment, there was evidence of recovery of populations of chironomids, mayflies, and *Daphnia*. On the 20th July, 90 days after treatment, populations of chironomids, mayflies, and *Daphnia* had fully recovered from the effects of the treatment and their numbers were similar to those in the control ponds.

Various indirect effects of the treatment occurred that could not be predicted on the basis of laboratory studies. Firstly, there was an increase in the number of Diaptomidae in treated ponds during the period 20–90 days after treatment (Fig. 6). This effect is very similar to that reported by Hurlbert et al. [5], who investigated the ecological effects of treatments of the organophosphorus insecticide chlorpyrifos in shallow (24 cm) ponds near Bakersfield, California. An observed increase in *Diaptomus* populations after treatment was thought to be attributable to mortality of predatory *Cyclops* populations.

Secondly, MEP treatment had a significant effect ($P < 0.05$) on the growth of rainbow trout. The maximum concentration of MEP to which the trout were exposed in this field experiment was 90 μg l^{-1} whereas in the laboratory the growth of rainbow trout was not affected when they were exposed continuously for 28 days to concentrations up to 200 μg l^{-1} [80]. It was concluded that growth was depressed by the effect of MEP on the food supply of trout, i.e. the aquatic invertebrates.

Finally, in one of the three treated ponds there was a bloom of filamentous algae which subsequently collapsed, leading to severe depression of the concentration of dissolved oxygen and mortality of some of the rainbow trout, 37 days after treatment. These effects may have been caused by mortality of daphnids and mayfly larvae that feed on filamentous algae and thus exert a controlling effect on their rate of growth. Since effects on algae and dissolved oxygen concentrations occurred in only one of the three treated ponds they cannot be attributed with any certainty to the MEP treatment. In the other two treated ponds there were relatively fewer filamentous algae and therefore there was no possibility that these effects could occur. It may be concluded that considerably more than three replicates would be needed to study secondary effects of this nature, i.e. where secondary biological effects may or may not occur depending on the nature of the pond and the kind of organisms present.

Conclusions

Purpose-built outdoor ponds can play a useful role in experimental ecotoxicology, intermediate between laboratory testing and environmental monitoring. Carefully constructed and managed systems can closely simulate the natural environment while at the same time they provide opportunity for replication of treatments.

Replicated experiments in outdoor ponds have proved to be of value in obtaining information concerning the environmental fate and biological effects of chemicals in freshwater. Dispersion and degradation in the field can be underestimated by extrapolation from simple laboratory experiments. Interactions between dispersive fluxes and degradation processes can be important and they need to be studied in the field.

Experiments in ponds may be useful in evaluating toxicity of chemicals to a much wider range of organisms than is possible in the laboratory. It is also possible to study interactions between chemicals, aquatic organisms, and their environment. Such interactions can cause biological effects that may be more important to the integrity of freshwater ecosystems than direct toxicity.

References

1. Hall, D.J., Cooper, W.E., Werner, E.E.: Limnol. Oceanogr. *15*, 839 (1970)
2. O'Brien, W.J., De Noyelles, F. Jr.: Hydrobiologia *44*, 105 (1974)
3. DeNoyelles, F. Jr., O'Brien, W.J.: Arch. Hydrobiol. *84*, 137 (1978)
4. Krishnamoorthi, K.P., Abdulappa, M.K., Sarkar, R., Siddiqi, R.H.: Water Res. *9*, 269 (1975)
5. Hurlbert, S.H., Mulla, M.S., Willson, H.R.: Ecol. Monogr. *42*, 269 (1972)
6. Apperson, C.S., Schaefer, C.H., Colwell, A.E., Werner, G.H., Anderson, N.L., Dupras, E.F. Jr., Longanecker, D.R.: J. Econ. Entomol. *71*, 521 (1978)
7. Beynon, K.I., Edwards, M.J., Thompson, A.R., Edwards, C.A.: Pestic. Sci. *2*, 5 (1971)
8. Boyle, T.P.: Environ. Pollut., Ser. A *21*, 35 (1980)
9. Boyle, T.P.: Proc. Conf. "Aquatic plants, lake management and ecosystem consequences of lake harvesting," pp. 269–283, University of Wisconsin-Madison, February 14–16, 1979
10. Bridges, W.R.: Trans. Am. Fish. Soc. *90*, 332 (1961)
11. Bridges, W.R., Kallman, B.J., Andrews, A.K.: Trans. Amer. Fish Soc. *92*, 421 (1963)
12. Corbet, R.L., Muir, D.C.G., Webster, G.R.B.: Chemosphere *12*, 523 (1983)
13. DeNoyelles, F., Kettle, W.D., Sinn, D.E.: Ecology *63*, 1285 (1982)
14. Elner, J.K., Wildish, D.J.: Can. Tech. Rep. Fish. Aquat. Sci. No. 933, Jan 1981
15. Hughes, D.N., Boyer, M.G., Papst, M.H., Fowle, C.D., Rees, G.A.V., Baulu, P.: Arch. Environm. Contam. Toxicol. *9*, 269 (1980)
16. Hurlbert, S.H., Mulla, M.S., Keith, J.O., Westlake, W.E., Düsch, M.E.: J. Econ. Entomol. *63*, 43 (1970)
17. Kudo, A., Garrec, J.P.: Regulatory Toxicology and Pharmacology *3*, 189 (1983)
18. Lorio, W.J., Jenkins, J.H., Huish, M.T.: Trans. Am. Fish. Soc. *105*, 695 (1976)
19. Macek, K.J., Walsh, D.F., Hogan, J.W., Holz, D.D.: Trans. Am. Fish. Soc. *101*, 420 (1972)
20. Mathis, M.J., Cummings, T.E., Gower, M., Taylor, M., King, C.: Hydrobiologia *67*, 197 (1979)
21. Mauck, W.L., Mayer, F.L. Jr., Holz, D.D.: Bull. Environ. Contam. Toxicol. *16*, 1 (1976)
22. Muir, D.C.G., Grift, N.P., Lockhart, W.L.: Can. J. Fish. Aquat. Sci. *39*, 1273 (1982)
23. Mulla, M.S., Darwazeh, H.A.: Proc. Calif. Mosq. Assoc. *43*, 164 (1975)
24. Papst, M.H., Boyer, M.G.: Hydrobiologia *69*, 245 (1980)
25. Rawn, G.P., Webster, G.R.B., Muir, D.C.G.: J. Environ. Sci. Health, *B17*, 463 (1982)
26. Sanders, H.O., Walsh, D.F., Campbell, R.S.: Technical Papers of the U.S. Fish and Wildlife Service No. 104, Washington, D.C., 1981
27. Sayler, G.S., Perkins, R.E., Sherrill, T.W., Perkins, B.K., Reid, M.C., Shields, M.S., Kong, H.L., Davis, J.W.: Appl. Environ. Microbiol. *46*, 211 (1983)
28. Scott, B.S., Dutka, B.J., Sherry, J.P., Glooschenko, V., Wade, P.J., Taylor, W.D., Nagy, E., Snow, N.B., Carlisle, D.B.: Scientific Series No. 130, Canada Centre for Inland Waters, Burlington, Ontario 1982
29. Troeger, W.W.: Proc. Okla. Acad. Sci. *61*, 79 (1981)
30. Tsushimoto, G., Matsumura, F., Sago, R.: Environ. Toxicol. Chem. *1*, 61 (1982)
31. Van Valiu, C.C.: Advancement in Chemistry Series. Organic pesticides in the environment, pp. 271–279 (1966)
32. Adams, W.J., Kimerle, R.A., Heidolph, B.B., Michael, P.R.: In Bishop, W.E., Cardwell, R.D., Heidolph, B.B. (eds.). Aquatic Toxicology and Hazard Assessment: pp. 367–385 Sixth Symposium, ASTM STP 802, Philadelphia 1983
33. Muir, D.C.G., Grift, N.P., Lockhart, W.L.: Environ. Toxicol. Chem. *1*, 113 (1982)
34. Hansen, S.R., Garton, R.R.: Can. J. Fish. Aquat. Sci. *39*, 1273 (1982)
35. Baughman, G.L., Burns, L.A.: In Hutzinger, O. (ed.). The Handbook of Environmental Chemistry, Volume 2, Part A, Reactions and Processes, pp. 1–17; Springer-Verlag, Berlin, Heidelberg 1980
36. Smith, H., Mabey, W.R., Bohonos, N., Holt, B.R., Lee, S.S., Chou, T.-W., Bomberger, D.C., Mill, T.: "Environmental pathways of selected chemicals in fresh water systems. Part I. Background and experimental procedures," EPA-600/7-77-113, (1977), and "Part II. Laboratory studies," EPA-600/7-78-074, (1978). National Technical Information Service, Springfield, Virginia 22161

37. Wolff, C.J.M., van der Heijde, H.B.: Chemosphere *11*, 103 (1982)
38. Zepp, R.G., Cline, D.M.: Environ. Sci. Technol. *11*, 359 (1977)
39. Paris, D.F., Steen, W.C., Baughman, G.L., Barnett, J.T. Jr.: Applied Environ. Microbiol. *41*, 603 (1981)
40. Pritchard, P.H.: In Richard A. Conway (ed.) Environmental Risk Analysis, pp. 257–342; Van Nostrand Reinhold, New York 1982
41. Hurlbert, S.H.: Res. Rev. *57*, 81 (1975)
42. Caspers, N., Hamburger, B.: Z. Wasser-Abwasser-Forsch. *16*, 205 (1983)
43. Kennedy, H.D., Eller, L.L., Walsh, D.F.: Technical Papers of the U.S. Fish and Wildlife Service No. 53, Washington, D.C. 1970
44. Robinson-Wilson, E.F., Boyle, T.P., Petty, J.D.: In Bishop, W.E., Cardwell, R.D., Heidolph, B.B. (eds.) Aquatic Toxicology and Hazard Assessment pp. 239–251: Sixth Symposium, ASTM STP 802, Philadelphia 1983
45. Brooker, M.P., Edwards, R.W.: J. Inst. Fish. Mgmt. *4*, 102 (1973)
46. Brooker, M.P., Edwards, R.W.: Freshwat. Biol. *3*, 157 (1973)
47. Brooker, M.P., Edwards, R.W.: Freshwat. Biol. *3*, 375 (1973)
48. Dickson, K.L., Maki, A.W., Cairns, J. Jr. (eds.): Analyzing the Hazard Evaluation Process. Am. Fish. Soc., Washington, D.C. 1979
49. Neely, W.B., Blau, G.E., Alfrey, T. Jr.: Environ. Sci. Technol. *10*, 72 (1976)
50. Neely, W.B.: Chemicals in the environment: distribution, transport, fate analysis. Marcel Dekker, New York 1980
51. Burns, L.A., Cline, D.M., Lassiter, R.R., "EXAMS: an exposure analysis modelling system," Preliminary draft document available through R.R. Lassiter, United States Environmental Protection Agency, Environmental Research Laboratory, College Station Road, Athens, GA 30613, February 1980
52. Mackay, D., Paterson, S.: Environ. Sci. Technol. *16*, 654A (1982)
53. Mackay, D., Yeun, A.T.K.: Environ. Sci. Technol. *17*, 211 (1983)
54. Rodgers, H. Jr., Dickson, K.L., Saleh, F.Y., Staples, C.A.: Environ. Toxicol. Chem. *2*, 155 (1983)
55. Schauerte, W., Lay, J.P., Klein, W., Korte, F.: Ecotoxicol. Environ. Saf. *6*, 560 (1982)
56. Lyman, W.J., Reehl, W.F., Rosenblatt, D.H.: Handbook of chemical property estimation methods, McGraw-Hill, New York 1982
57. Karickhoff, S.W.: Chemosphere *10*, 833 (1981)
58. Chiou, C.T., Porter, P.E., Schmedding, D.W.: Environ. Sci. Technol. *17*, 227 (1983)
59. Voice, Th.C., Weber, W.J. Jr.: Water Res. *17*, 1433 (1983)
60. Weber, W.J. Jr., Voice, Th.C., Pirbazari, M., Hunt, G.E., Ulanoff, D.M.: Water Res. *17*, 1443 (1983)
61. O'Connor, D.J., Connolly, J.P.: Water Res. *14*, 1517 (1980)
62. Callahan, M.A., Slimak, M.W., Gabel, N.W., May, I.R., Fowler, C.F., Freed, J.R., Jennings, P., Durfee, R.L., Whitmore, F.C., Maestri, B., Mabey, W.R., Holt, B.R., Gould, C.: Water related environmental fate of 129 priority pollutants Vol. II, EPA-440/4-79-029b, December 1979
63. OECD (Organisation for Economic Co-operation and Development), OECD guidelines for testing of chemicals, OECD, Paris 1981
64. Mabey, W., Mill, T.: Phys. Chem. Ref. Data *7*, 383 (1978)
65. Perdue, E.M., Wolfe, N.L.: Environ. Sci, Technol. *16*, 847 (1982)
66. Zepp, R.G.: Environ. Sci. Technol. *12*, 327 (1978)
67. ECETOC (European Chemical Industry Ecology and Toxicology Centre), The phototransformation of chemicals in water: results of a ring test. Technical Report No. 12, ECETOC, Brussels
68. Mo, T., Green, A.E.S.: Photochem. Photobiol. *20*, 483 (1974)
69. Crossland, N.O., Wolff, C.J.M.: Environ. Toxicol. Chem. *in press*
70. Wolff, C.J.M., Crossland, N.O.: Environ. Toxicol. Chem. *in press*
71. Zepp, R.G., Wolfe, N.L., Baughman, G.L., Hollis, R.C.: Nature *267*, 421 (1977)
72. Wolff, C.J.M., Halmans, M.T.H., van der Heijde, H.B.: Chemosphere *10*, 59 (1981)
73. Momzikoff, A., Santus, R., Giraud, M.: Mar. Chem. *11*, 1 (1983)
74. Cooper, W.J., Zika, R.G.: Science *222*, 711 (1983)

75. Garland, J.A., Elzerman, A.W., Penkett, S.A.: J. Geophys. Res. *85*, 7488 (1980)
76. Mill, T., Hendry, D.G., Richardson, H.: Science *207*, 886 (1980)
77. Miller, G.C., Zisook, R., Zepp, R.G.: J. Agric. Food Chem. *28*, 1053 (1980)
78. Crossland, N.O., Bennett, D.: Ecotoxicol. Environ. Saf. *8*, 471 (1984)
79. Butcher, J.R., Boyer, M.G., Fowle, C.D.: Proc. 10th Can. Symp. Water Poll. Research, pp. 33–41 (1975)
80. Crossland, N.O.: Ecotoxicol. Environ. Saf. *8*, 482 (1984)
81. Neely, W.B.: Environ. Sci. Technol. *13*, 1506 (1979)
82. Jessiman, B.J., Qadri, S.H.: Ecotoxicol. Environ. Saf. *7*, 295 (1983)
83. Spacie, A., Landrum, P.F., Leversee, G.J.: Ecotoxicol. Environ. Saf. *7*, 330 (1983)
84. Oliver, B.G., Niimi, A.J.: Environ. Sci. Technol. *17*, 287 (1983)
85. Thomann, R.V., Connolly, J.P.: Environ. Sci. Technol. *18*, 65 (1984)
86. Garten, C.T. Jr., Trabatha, J.R.: Environ. Sci. Technol. *17*, 590 (1983)
87. Banerjee, S., Sugatt, R.H., O'Grady, D.P.: Environ. Sci. Technol. *18*, 79 (1984)
88. Ellgehausen, H., Guth, J.A., Esser, H.O.: Ecotoxicol. Environ. Saf. *4*, 134 (1980)
89. Esser, H.O., Moser, P.: Ecotoxicol. Environ. Saf. *6*, 131 (1982)

Hydrolysis of Organic Chemicals

Theodore Mill

Chemistry Laboratory, SRI
Menlo Park, CA 94025, USA

William Mabey

Kennedy/Jenks Engineers
San Francisco, CA 94103, USA

Summary

Hydrolysis of organic compounds in water usually depends on the concentrations of protons or hydroxide ion or both, as well as water. Mechanisms of hydrolysis are well-described by classical concepts of substitution or addition by HO^- or H_2O at sp^3 or sp^2 carbon. This process may or may not involve carbenium ions, depending on the structure of the carbon skeleton and the leaving group. Thus the rate limiting step may be kinetically first or second order. Generally substitution at sp^3 carbon does not require acid catalysis except for leaving groups with oxygen or sulfur such as found in acetals, ketals or epoxides or their thio analogs.

Addition-elimination reactions at sp^2 carbonyl groups usually involve acid catalysis as well as direct reaction with water and HO^- ion and these reactions almost always are kinetically second order.

In pure water, hydrolysis of neutral organic compounds follow the general rate expression

$$- dC/dt = k_A[H^+][C] + k_B[HO^-][C] + k_N[C],$$

where k_A, k_B, and k_N are acid, base and neutral rate constants, respectively, and where values of any two rate constants may be zero depending on the structure of the compound.

Kinetically more complicated situations are found when hydrolysis reactions also are catalysed by general acid or bases, metal ions or by ionic strength effects. Quantitative assessment of these effects in natural waters reveals that in almost all cases these added catalysis terms are too small to be significant. A similar conclusion applies to common-ion effects which may occur with benzylic halides in marine environments.

Structure activity relationships (SARs) for hydrolysis are based on Hammett or Taft relationships. Relevant SAR parameters for several classes of carbonyl compounds are reviewed.

Introduction

The enormous worldwide production of synthetic organic chemicals, now amounting to about 400 billion kilograms per year, inevitably leads to adventitious but significant introduction of many hundreds of different organic chemicals into major streams, rivers and coastal waters in most industrial countries. Many of these chemicals react with water or components of it to form new chemical structures with halflives that vary widely from a few seconds to a few thousand years.

Although many of the least reactive compounds will probably be transformed or lost by competing processes more quickly, hydrolysis is a major transformation process for many kinds of organic chemicals in water and the rates and products of these processes are of considerable concern to environmental chemists and those involved in regulating production and disposal of organic chemicals in the environment. Transformation of chemical structures to new ones may mitigate or exacerbate the potential toxic effects of these chemicals on the biological communities. Therefore, it is of great importance that we are able to understand and predict how quickly chemicals transform by this process and the likely products of the transformation.

This chapter reviews our current understanding of hydrolysis processes, effects of chemical structure on the rates of these reactions and effects of environmental properties of fresh and marine waters on hydrolysis. In addition the review also provides a list of rate constants characteristic for hydrolysis of many important classes of organic compounds. We also indicate some recommended laboratory methods for measuring rate constants for hydrolysis of organic compounds in water.

Hydrolysis as treated in here refers specifically to introduction of water into an organic molecule accompanied by cleavage of another chemical bond, excluding cleavage of carbon-carbon or carbon-oxygen double bonds; these latter processes, often referred to as hydration, are important but are not considered further. Simple examples of hydrolysis reactions include conversion of alkyl halides to alcohols, conversion of esters to acids and epoxides to diols. A useful generalization is that hydrolysis may be important in any molecule which has carbon or carbon-oxygen or carbon-nitrogen double bonds linked to electronegative atoms or groups through σ-bonds.

Hydrolysis reactions have been studied in the laboratory for well over 100 years. As a result an enormous literature has accumulated with very substantial numbers of rate constants, structure activity relationships and useful general information about the reaction [1]. Many reviews have been written concerning specific aspects of hydrolysis; a comprehensive review of kinetic constants for hydrolysis of organic compounds in water was prepared by Mabey and Mill [2] and we recommend it for supplementary kinetic information. Valuable discussions of detailed mechanism can be found in reviews by Streitweiser [3], Thornton [4], and Bamford and Tipper [5].

Mechanisms and Kinetics of Hydrolysis

Kinetic Relations

The rate equation for hydrolysis of many chemicals (RX) usually can be expressed as

$$d[RX]/dt = k_h[RX] = (k_A[H^+] + k_N + k_B[OH^-])[RX], \qquad (1)$$

where k_B, k_A, and k_N are process rate constants for the acid- and base-catalyzed neutral processes, respectively. Generally, the concentration of water is nearly constant and much greater than the chemical RX, therefore $k'_N[H_2O]$ is also a constant, k_N. The pseudo first-order rate constant k_h is the observed or estimated rate constant for hydrolysis at a specific and constant pH and temperature. Equation (1) throughout assumes that the individual rate processes for the acid, base, and neutral hydrolyses are each first-order in chemical. Since this is usually the case

$$k_h = k_A[H^+] + k_N + k_B[OH^-] \qquad (2)$$

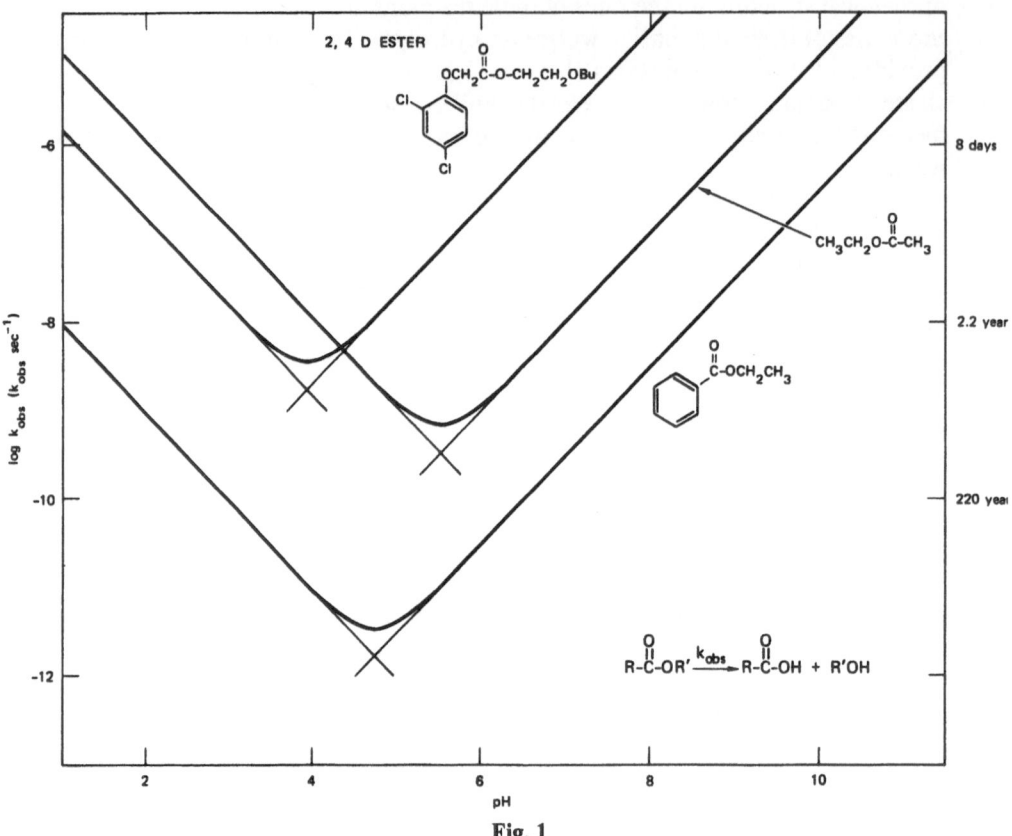

Fig. 1

From the autoprotolysis water equilibrium [Eq. (3)] Eq. (2) may be rewritten as Eq. (4)

$$K_w = \frac{[H^+][OH^-]}{[H_2O]},$$ (3)

$$k_h = k_A[H^+] + k_N + K_w k_B/[H^+].$$ (4)

From Eq. (4), it is evident how pH effects the overall rate; at high or low pH the first or last term is usually dominant, whereas at pH 7 k_N often is most important. However, the detailed relationship and rate depends on the specific values of k_B, k_A, and k_N. At any fixed pH the overall rate process is pseudo-first order and the half-life of the chemical is independent of its concentration

$$t_{1/2} = \ln 2/k_h.$$ (5)

The relation of the rate constant to pH is conveniently expressed graphically in the form of $\log k_h$ versus pH in which the three equations one each for the acid, base and neutral processes are plotted individually. Curves which are obtained in this way are especially useful for describing effects of acid or base on the rate of hydrolysis of a specific chemical. Figure 1 depicts a typical $\log k_h$ versus pH plot for esters which undergo acid, water, and base promoted hydrolyses. It is obvious from Eq. (4) that each curve in Fig. 1 is a composite of three straight lines (a) $\log k_h = \log k_A - pH$; (b) $\log k_h = k_N$, and (c) $\log k_h = \log(k_B K_w) + pH$ with slopes of -1, 0, and $+1$, respectively, [2].

Mechanisms of Hydrolysis

Hydrolyses reactions may be considered as subsets of the broader classes of substitution reactions on aliphatic and carboxylic acid derivatives. Substitution reactions are those in which group Y replaces group X on either sp^3 or sp^2 carbon as shown in Eqs. (6) and (7).

$$RX + Y^- \longrightarrow RY + X^-, \tag{6}$$

$$\overset{\displaystyle O}{\underset{\displaystyle \|}{R\overset{}{C}}}-X + Y^- \longrightarrow \overset{\displaystyle O}{\underset{\displaystyle \|}{R\overset{}{C}}}-Y + X^-. \tag{7}$$

Aliphatic substitution on saturated (sp^3) carbon involves displacement of one group X with its bonding electrons by another group usually having an unbounded electron pair [reaction (6)]. This reaction is an example of a Lewis acid-base reaction in which one base replaces another. The two major types of aliphatic substitution reactions in environmental situations are those involving neutral water or OH^- displacing on a neutral species.

Physical organic chemists have devoted 60 years to intensive study of substitutions, and have developed an extensive data base, mostly in organic solvents for these processes. A number of excellent reviews are available which the reader should consult for mechanistic details. A general overview of these reaction mechanisms is given by Lowry and Richardson [1]; more detailed reviews of substitution reactions may be found in Streitweiser [3], Thornton [4], Harris [6], and Bentley and Schleyer [7]. Below we discuss some of the general features of nucleophilic substitution both at sp^3 and sp^2 carbons and the relation of mechanism to kinetics and structure activity relationships (SARs).

Aliphatic Substitution

Fifty years ago, Hughes and Ingold postulated that two limiting processes or some combination of these described all aliphatic substitutions [8]. According to the Hughes-Ingold theory the two limiting processes involve either (1) nucleophilic displacement by Y on X to produce a new $C-Y$ bond with release of X^- a process labeled the S_N2 mechanism or (2) rate controlling heterolytic fission of the $C-X$ bond to form a carbenium ion and anion pair which captures nucleophile Y to form a new $C-Y$ bond

$$Y^- + -\overset{\diagdown}{\underset{\diagup}{C}}-X \longrightarrow Y-\overset{\diagup}{\underset{\diagdown}{C}}- + X^- \quad S_N2, \tag{6}$$

$$-\overset{\diagdown}{\underset{\diagup}{C}}-X \longrightarrow -\overset{|}{\underset{|}{C}}{}^+X^- \overset{Y^-}{\longrightarrow} -\overset{\diagdown}{\underset{\diagup}{C}}-Y + X^- \quad S_N1. \tag{8}$$

This latter process is labeled S_N1. This mechanistic framework is still valid although we now recognize that these two extreme processes are found only in small fraction of the universe of substitution reactions actually observed; in many cases nucleophilic substitution involves a combination of partial heterolytic cleavage assisted by the nucleophile (often the solvent). Substitution reactions in which solvent itself is the nucleophile are called solvolysis reactions. The term is merely descriptive and is not indicative of the mechanism of the process which might be either S_N2, S_N1 or some combination of these.

Detailed arguments concerning the participation by ion pairs in a particular substitution are complex, and will not be repeated here. The principle concern for mechanism is two-fold: the S_N1 process usually leads to unimolecular kinetics, whereas the S_N2 process gives bimolecular kinetics. Moreover, successful correlation of structure with reactivity requires that, within a class of compounds undergoing the same type of reaction, the detailed mechanism of the process be similar from one compound to another. Failure to observe this requirement is the most common cause for failure of structure activity relationships.

Some simple rules generally are useful in thinking about which process is most likely to be important. The S_N1 process occurs in systems in which the incipient carbonium ion is highly stabilized by electron-donating or conjugating substituents. Thus tertiary alkyl halides, α-haloethers, benzyl halides, and alkyl halides react largely by S_N1 processes reactions with the exception of bridgehead halides where the incipient and planar carbonium ion is sterically constrained by the bridge structure. Even in the case of t-butyl halides there is still significant nucleophilic assistance by solvent [3, 7]. The S_N2 process, on the other hand, occurs at sterically unhindered primary carbons, where stabilization of positive charge is also least accommodated. Thus n-alkyl halides, tosylates or sulfonates exhibit bimolecular displacement with little or no evidence for heterolytic cleavage in the rate controlling step [1, 3].

The nature of the leaving group X, also will strongly influence the rate at which a nucleophile Y can displace it. Thus halides or anions of other strong acids comprise the bulk of X groups which are displaced readily in water. Groups such as OH or NH_2 which are very weakly acidic are not readily displaced by H_2O or OH^-.

Hydrolysis of Carboxylic Acid Derivatives

Mechanisms of hydrolysis of carbonyl compounds are of great importance not only because of the widespread utility of carbonyl derivatives but because of the close connection to biological processes involving enzymatic hydrolysis. Therefore, a great deal of attention has been directed to the mechanisms of formation and cleavage of carboxylic acid derivatives. The diversity of mechanisms exhibited among these compounds is large and beyond the scope of this review. Again, we suggest the reader consult general reviews such as that of Lowry and Richardson [1] or more specialized reviews such as Patai [9], Fife [10], Jencks [11].

Esters, anhydrides, acyl halides, and amides are important carboxylic acid derivatives and these structures can be readily converted to the corresponding acid in the presence of either acid, base or both. The detailed mechanisms for hydrolyses follow a number of pathways most of which involve initial addition of a nucleophile such as water or OH^- to form a tetrahedral intermediate followed by loss of the other substituent to form the carboxylic acid or anion [12]

$$\underset{\|}{\overset{O}{RC}}-X + Y^- \rightleftharpoons \underset{\underset{Y}{|}}{\overset{O}{\underset{|}{RC}}}-X \longrightarrow RCY + X^- \qquad (9)$$

$$(Y = H_2O \quad \text{or} \quad OH^-).$$

Thus addition-elimination is the carbonyl counterpart of the S_N2 process at saturated carbon. The exact counterpart for an S_N1 process in carbonyl systems is rarely observed; S_N1 acid catalyzed removal of leaving group X in acyl halides has been described [13]. We should also note that aliphatic substitution (S_N2) can be observed when the carboxylic acid itself is highly electronegative and the forms a stabilized anion. For example in primary or secondary alkyl trifluoroacetates, substitution occurs at the saturated carbon by displacement of the trifluoroacetate moiety

$$Y^- + R\underset{\|}{\overset{O}{O}}CCF_3 \longrightarrow YR + CF_3CO_2^-. \qquad (10)$$

Catalysis by acids and bases is a characteristic feature of carboxylic acid chemistry. In most cases acid catalysis involves initial protonation of the carbonyl oxygen [12]. This process favors addition by a nucleophile such as water (OH^- usually is in too low concentration to be important) to an otherwise relatively unactivated carbonyl. Esters and amides are examples of carbonyl derivatives in which acid catalysis is important; acyl halides are examples in which it is not important.

Both OH^- and H_2O act as nucleophiles on unactivated carbonyl derivatives. Substitution by hard base OH^- generally is much faster than by soft base H_2O, typically by a factor of $10^5 - 10^6$ with the result that water promoted hydrolysis is difficult to observe in many systems. However, for some types of esters, water-promoted hydrolysis is the dominant process in the pH range 6–8 and unfortunately the data base for these reactions is notably lacking [2, 5].

In acidic hydrolysis of both esters and amides (and related derivatives) protonation on carbonyl oxygen is the dominant process and unlike the S_N2 process, displacement of poor leaving groups such as RO^- or RNH^- can now be effected by H_2O. This process is fully reversible at each step; however, the last step forms carboxylate anion which effectively removes it from the reaction scheme

$$\underset{\underset{OH_2^+}{|}}{\overset{HO}{\underset{|}{RC}}}-OR' \longrightarrow R\overset{O}{\overset{\|}{C}}OR' + H_2O + H^+, \qquad (11)$$

$$\underset{\underset{OH_2^+}{|}}{\overset{HO}{\underset{|}{RC}}}-OR' \longrightarrow R\overset{O}{\overset{\|}{C}}O^- + R'OH. \qquad (12)$$

It was recognized many years ago by Ingold [14] that, while the base catalyzed hydrolysis of esters is subject to strong acceleration by electron attracting groups such as chlorine, acid catalyzed hydrolysis is relatively insensitive to polar effects [14–16]. This difference is the basis for Taft's important structure activity relationship for hydrolysis of aliphatic compounds [16]. DeTar and Tenpas [15] successful theoretical calculation of the steric effect in ester hydrolysis suggests that steric effects arise primarily from an enthalpy factor. Apparently steric conditions at the ester group and at the tetrahedral intermediate are relatively constant from one ester to another with the result that only small entropy effects are noted.

The Tetrahedral Addition Intermediate

Hydrolysis of carboxylic acid derivatives usually proceeds through a tetrahedral intermediate, the existence of which can be inferred from several kinds of evidence: (1) ^{18}O exchange (2) breaks in the pH-rate profile which are not due to ionization (3) breaks in buffer concentration versus rate curves and (4) breaks in reactivity rate relationships. Detailed kinetic analysis of the hydrolysis of carboxylic acid derivatives is given by Talbot [17]. The hydrolysis process may be written as shown below

$$\mathrm{R\overset{\overset{\displaystyle O}{\|}}{C}OR' + H_2O^*} \underset{2}{\overset{1}{\rightleftharpoons}} \mathrm{R\underset{\underset{\displaystyle {}^*OH}{|}}{\overset{\overset{\displaystyle OH}{|}}{C}}{-}OR'}, \tag{13}$$

$$\mathrm{R\underset{\underset{\displaystyle {}^*OH}{|}}{\overset{\overset{\displaystyle OH}{|}}{C}}{-}OR'} \xrightarrow{3} \mathrm{R\overset{\overset{\displaystyle O}{\|}}{C}{}^*OH + R'OH}, \tag{14}$$

where H_2O^* is labeled to illustrate oxygen exchange.

Exchange of ^{18}O into the intermediate in hydrolysis of carboxylic acid derivatives provides compelling evidence for the existence and lifetime of the tetrahedral intermediate for which Bender [12] has provided detailed and elegant studies. In addition to Eqs. (13) and (14) we can write

$$\mathrm{R\underset{\underset{\displaystyle {}^*OH}{|}}{\overset{\overset{\displaystyle OH}{|}}{C}}{-}OR} \underset{1'}{\overset{2'}{\rightleftharpoons}} \mathrm{R\overset{\overset{\displaystyle O^*}{\|}}{C}OR + H_2O} \tag{15}$$

which leads to oxygen exchange. The amount of ^{18}O recovered from unhydrolyzed derivatives reflects the partitioning of the tetrahedral intermediate, providing that this intermediate lies on the reaction pathway. In the above equations the ratio $k_{2'}/2k_3$ is the ratio of exchange to hydrolysis. As $k_{2'}$ increases, the ratio increases in the order amides > esters > anhydrides > acid halides. Thus the better the leaving group, the smaller the value of $k_2/2k_3$. Since the process shown by Eqs. (13)–(14) give an overall observed rate constant defined by

$$k_{obs} = k_1 k_3/(k_2 + k_3) = k_1/[(k_2/k_3) + 1].$$ (17)

Values of $k_2/k_3 \ll 1$ show that the addition step (k_1) must be rate determining. Examples of small ratios are the alkaline, neutral and acid hydrolyses of esters, acid hydrolyses of anhydrides and amides and neutral hydrolyses of acid chlorides, and anhydrides. Conversely when $k_2/k_3 \gg 1$ the rate determining step is breakdown of the tetrahedral intermediate. Examples of this limiting process include base promoted hydrolysis of amides. Detection of these intermediates by spectrometric techniques is quite difficult because of the short lifetimes, estimated at about 10^{-9} s or less, corresponding to about 6 kcal for the depth of the well in the reaction profile leading to products [1, 12, 17].

General Acid- and Base-Promoted Reactions

In addition to H_3O^+, HO^-, and H_2O, other chemical species may act as acids or bases to promote hydrolysis.[1] In the presence of general acid or base Z, the first order rate constant, k_h, for loss of chemical becomes

$$k_h = k_A[H^+] + k_N + k_B[OH] + k_z[Z],$$ (18)

where k_z and [Z] are the rate constant and concentration, respectively of Z; k_N, k_A, and k_B are the neutral, acid-promoted, and base-promoted rate constants, respectively [2]. The reaction of Z is referred to as a general acid- or general base-promoted process (depending on whether Z is an acid or base), and these reactions are distinguished from specific acid (HO_3^+) or specific base (HO^-) promoted processes. For example a general base-promoted process is aliphatic substitution in which Y displaces leaving group X from RX; if Y is itself a good leaving group the reaction product often is the same as formed in the uncatalyzed process

$$R-X + Y^- \longrightarrow R-Y + X^-,$$ (19)

$$R-Y + H_2O \longrightarrow ROH + HY.$$ (20)

General acid- or base-promoted processes are most often of concern in laboratory studies conducted with high buffer concentrations, where "buffer catalysis" effects can introduce artifacts into hydrolysis data [18].

Evaluation of the importance of general base-promoted processes must include the nucleophilic displacement reactions above. Relative nucleophilicities of several anionic species have been measured by Swain and Scott [19], where a nucleophilicity constant, n, is defined by Eq. (21),

$$\log \left(\frac{k_Y}{k_{H_2O}} \right) = sn,$$ (21)

where k_Y and k_{H_2O} are second order rate constants for displacement of X by a nucleophile Y or water, respectively, and s is a parameter for the susceptibility of

---

1 We use the term catalyzed-process when the overall stoichiometry of the process does not change the catalyst concentration.

Table 1. Relative nucleophilicities of selected anions

Nucleophile	n^a	k_{rel}	Y^* (M)^b
SO_3^{2-}	5.1	1×10^5	6×10^{-4}
HS^-	5.1	1×10^5	6×10^{-4}
I^-	5.0	1×10^5	6×10^{-4}
HO^-	4.2	2×10^3	3×10^{-3}
Br^-	3.9	8×10^3	7×10^{-3}
$HPO_4^=$	3.8	6×10^3	9×10^{-3}
HCO_3^-	3.8	6×10^3	9×10^{-3}
Cl^-	3.0	1×10^3	6×10^{-2}
SO_4^{2-}	2.5	3×10^2	2×10^{-1}
H_2O	0.0	1	–
ClO_4^-	<0	–	–

[a] n is nucleophilicity parameter in equation $\log(k/k_{H_2O}) = sn$, where $s = 1$ for methyl bromide in water and k is rate constant for nucleophilic displacement by anion from Swain and Scott [19]

[b] Y^* is concentration at which $55 k_{H_2O} = k_{rel}Y$

the reaction to nucleophilic reactions. Values of n were determined taking $n = 0$ for water and $s = 1$ for methyl bromide as the reactant

$$CH_3Br + H_2O \xrightarrow{k_{H_2O}} CH_3OH + HBr, \qquad (22)$$

$$CH_3Br + Y^- \xrightarrow{k_Y} CH_3-Y + Br^-. \qquad (23)$$

Table 1 lists values of n for some anions found in natural waters, values of k_y relative to water, and the concentrations of anions at which the nucleophilic displacement by the anion is competitive with water in the reaction with methyl bromide. Except for Cl^- in seawater (where $[Cl^-] = 0.3$ M), anion concentrations in natural waters have only marginal effects on hydrolysis rate constants of most chemicals.

Neighboring Group Reactions

A classical example of a neighboring group (NG) reaction is the pH-independent hydrolysis of bis(2-chloroethyl)sulfide (mustard) [20]

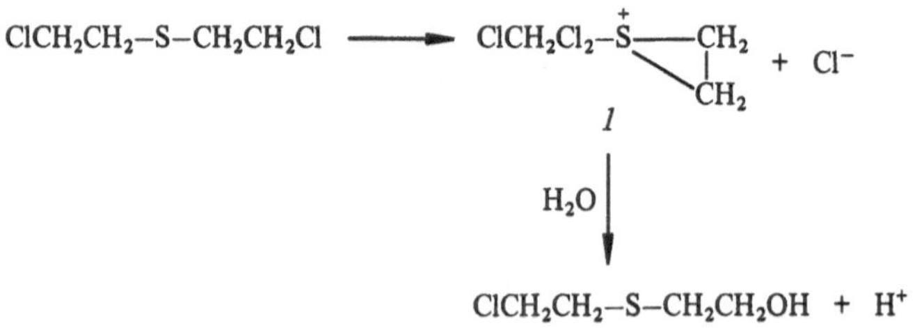

$$ClCH_2CH_2-S-CH_2CH_2Cl \longrightarrow \overset{+}{ClCH_2Cl_2-S} \overset{CH_2}{\underset{CH_2}{\diagdown}} + Cl^- \qquad (24)$$

$$1$$

$$H_2O \downarrow$$

$$ClCH_2CH_2-S-CH_2CH_2OH + H^+ \qquad (25)$$

Extensive studies on mustard have shown that the intramolecular process that produces the sulfonium ion is the rate determining step, and is orders of magnitude faster than hydrolysis resulting from simple displacement/solvolysis reaction with water. Most NG processes involving non-ionizable groups show pH-independent rates. These reactions typically involve 3-, 5-, or 6-membered-ring intermediates in which a nucleophile (N) assists in displacing a leaving group (L),

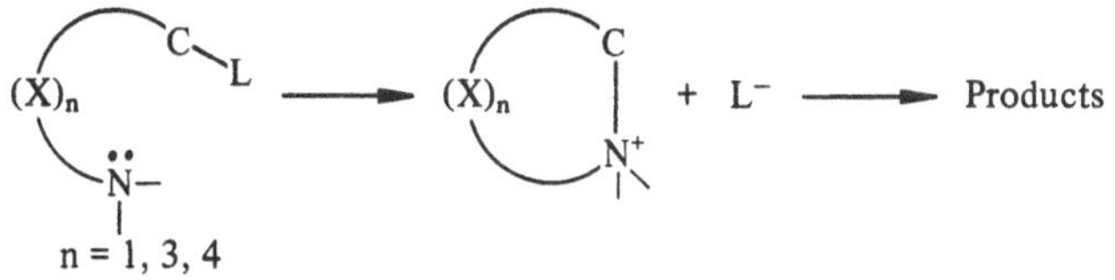

$$
\begin{array}{ccc}
\underset{n\,=\,1,\,3,\,4}{(X)_n} \overset{C-L}{\underset{\overset{|}{N-}}{}} & \longrightarrow & (X)_n \overset{C}{\underset{\overset{|}{N^+}}{}} \; + \; L^- \longrightarrow \text{Products}
\end{array} \qquad (26)
$$

where X are other atoms in the chain. Among non-ionized groups known to affect these reactions are halides, alkoxyl, amine, and carboxyl functions. Streitweiser [3] has reviewed a number of these reactions.

Although more rare, a NG process also can be pH dependent. A simple example is hydrolysis of S-n-propyl-o-(2-imidazoyl)thiobenzoate 2 [21]. The base-catalyzed process is 5×10^6 times greater than for the unsubstituted thiolbenzoate ester; the intermediate 3 was also independentally synthesized and compared spectrally with the reaction intermediate

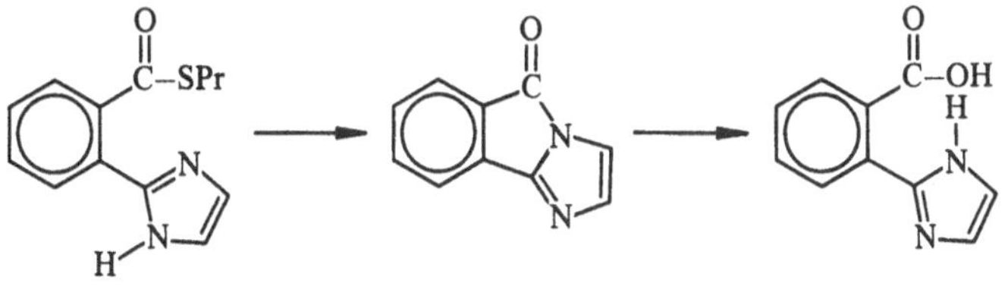

$$
\begin{array}{ccccc}
 & & & & \\
\textit{2} & & \textit{3} & & \textit{4}
\end{array} \qquad (27)
$$

Hydrolysis of Ionizable Chemicals

The pH-logk_h profiles of ionizable chemicals can be affected by both intra-molecular (or neighboring group) and electronic effects. An example of an electronic effect is found in the hydrolysis of N, N'-dimethyl-N-nitrosourea, 6 (Scheme 1)

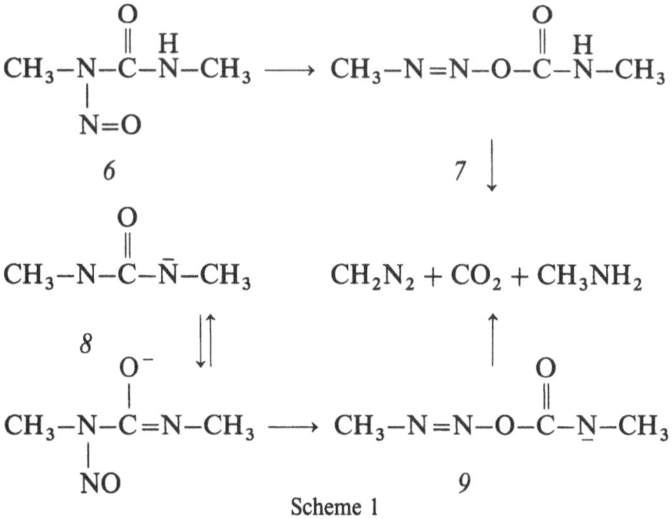

$$
\underset{\textit{6}}{\underset{\overset{|}{N=O}}{CH_3-N-\overset{\overset{\textstyle O}{\|}}{C}-\overset{\overset{\textstyle H}{}}{N}-CH_3}} \longrightarrow \underset{\textit{7}}{CH_3-N=N-O-\overset{\overset{\textstyle O}{\|}}{C}-\overset{\overset{\textstyle H}{}}{N}-CH_3}
$$

$$
\underset{\textit{8}}{CH_3-N-\overset{\overset{\textstyle O}{\|}}{C}-\bar{N}-CH_3} \qquad CH_2N_2 + CO_2 + CH_3NH_2
$$

$$
\underset{\textit{9}}{CH_3-N-\overset{\overset{\textstyle O^-}{|}}{C}=N-CH_3} \longrightarrow CH_3-N=N-O-\overset{\overset{\textstyle O}{\|}}{C}-\underset{}{N}-CH_3
$$

Scheme 1

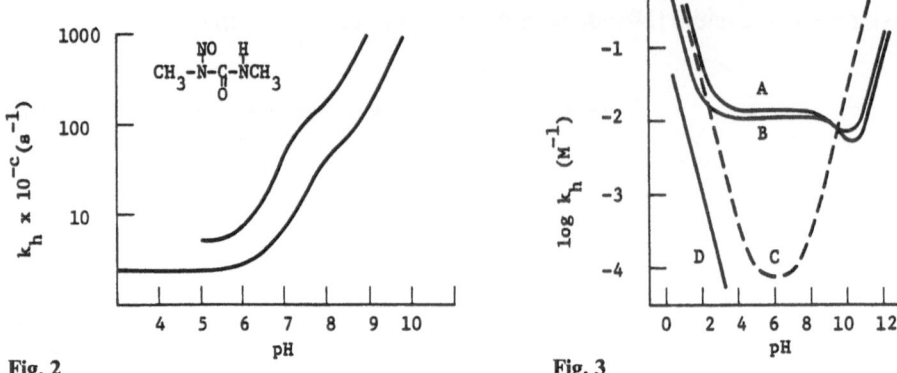

Fig. 2   Fig. 3

Fig. 2. k_h-pH profile for hydrolysis of N, N'-dimethyl-N-nitrosourea at 35 °C (lower curve) and 45 °C [22]

Fig. 3. Log k_h-pH profiles at 90 °C for hydrolysis of amides: Curve *A*, *o*-hydroxyphenylacetamide; *B*, *o*-hydroxyphenylpropionamide; *C*, γ-hydroxybutyramide; *D*, phenylpropionamide [23]

Compound *6* has a pK$_a$ of 8.2 at 35 °C [22]. The pH-logk_h profile (Fig. 2) for this chemical shows that the hydrolysis is pH-independent up to about pH 6, where base hydrolysis becomes important. At about pH 8, however, a transition point in the pH-log k_h profile occurs, and the profile then straightness out again at a pH of about 9. This profile is reasonably explained by occurrence of a neutral hydrolysis process (rate constant k_N) in the region pH 3 to 6, base-promoted hydrolysis (rate constant k_B) of the un-ionized species in the region pH 6 to about 8, and the base-promoted hydrolysis of the rearranged urea anion 9 (rate constant k'_B) above pH 9.

The first-order hydrolysis rate constant is given

$$k_h = (k_N + k_B[OH^-]f_C + k'_B[OH^-]f'_C,\qquad(28)$$

where f_C and f'_C are the fraction of chemical in the un-ionized and ionized forms. Although the anion undergoes the base-promoted reaction slower than the un-ionized species ($k_B = 22.9\,M^{-1}s^{-1}$ and $k'_B = 6.5\,M^{-1}s^{-1}$ at 35 °C), the f'_C term dominates k_h at higher pHs. Similar hydrolytic behavior has been reported by the same authors [22] for 1,6-dimethyl-1,6-dinitrosourea which also shows the simple acid-promoted process for the unionized form at pH values below pH 3.

A more complex example that demonstrates effects of both ionization and NG participation on a pH-logk_h profile is shown by the hydrolysis of *o*-hydroxy substituted aromatic amides [23]. As shown in Fig. 3, hydrolyses of these amides exhibits acid-promoted hydrolysis below pH 3 and, for *o*-hydroxyphenylalkyl amides, a pH independent process extending from pH 3 to about pH 7. This neutral hydrolysis process is not due to reaction with water, but rather reaction between the hydroxy group and the amide function. At a pH of about 9, the hydrolysis rate of the O-hydroxy amides is reduced because the phenolate anion forms and apparently does not participate in a intramolecular process.

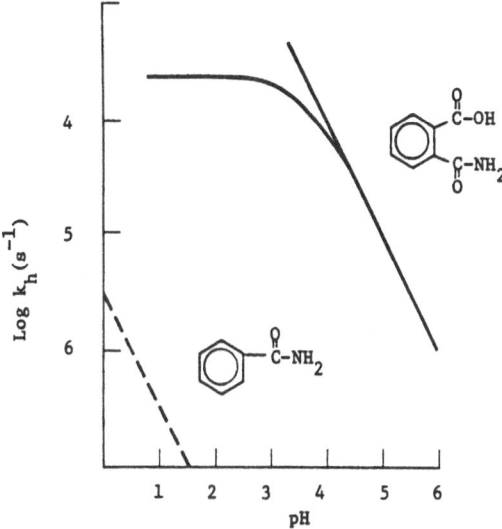

Fig. 4. $\log k_h$-pH profile at 47.3 °C for hydrolysis of phthalamic acid and benzamide [24]

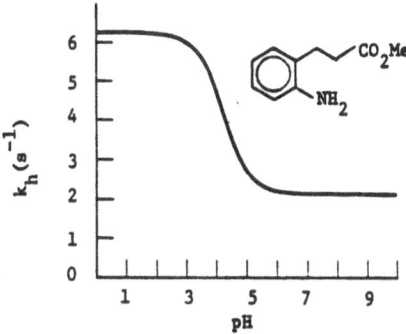

Fig. 5. $\log k_h$-pH profile for hydrolysis of methyl 3-(2-amino-phenyl) propionate at 39 °C [27]

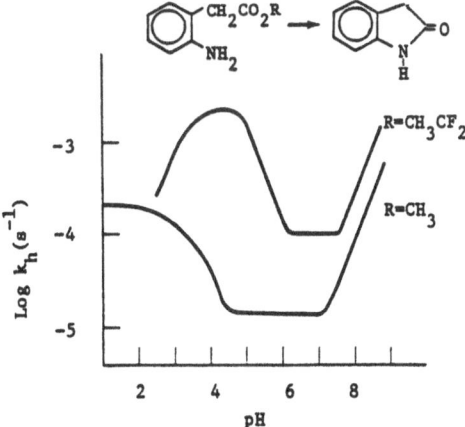

Fig. 6. $\log k_h$-pH profile for cyclization of alkyl (2-aminophenyl)acetates (R=Me at 50 °C, R=CH₃CF₂– at 30 °C) [28]

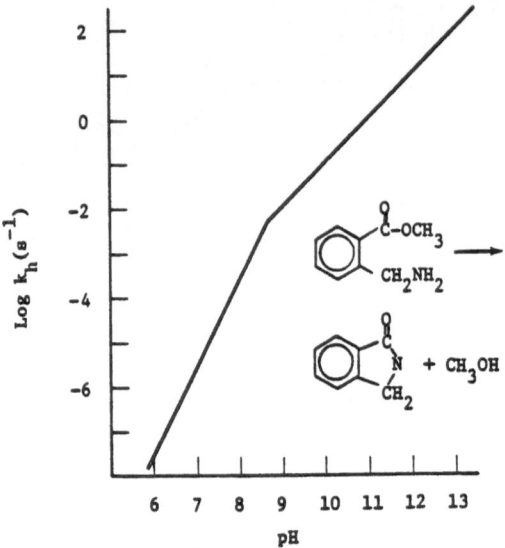

Fig. 7. Logk_h-pH profile for cyclization of methyl 2-aminoethylbenzoate to phthalimidine at 30 °C [21]

At about pH 10, however, base-promoted hydrolysis of the amide becomes important and dominates hydrolysis, as is evident by the slope of +1.

Another example of a profile involving NG reactions in amide hydrolyses has been shown by Bender et al. [24], who reported that the hydrolysis of phthalamic acid was pH-independent to about pH 3 because of intramolecular catalysis involving the carboxylic acid. Above pH 3, where the carboxylic acid ionizes ($pK_A = 3.7$), the hydrolysis rate slows with increasing pH, and it shows typical acid-promoted behavior with a slope of -1 in the pH-log k_h profile (see Fig. 4).

Amines also can act as neighboring groups. Examples of pH profiles for reaction of neighboring amines with esters are shown in Figs. 5 through 8, all of which involve cyclization of esters containing amines to form lactam intermediates. Jencks [25] has discussed these complex systems in terms of tetrahedral intermediates that are stabilized or destabilized depending on the solution pH and substrate structure (Scheme 2)

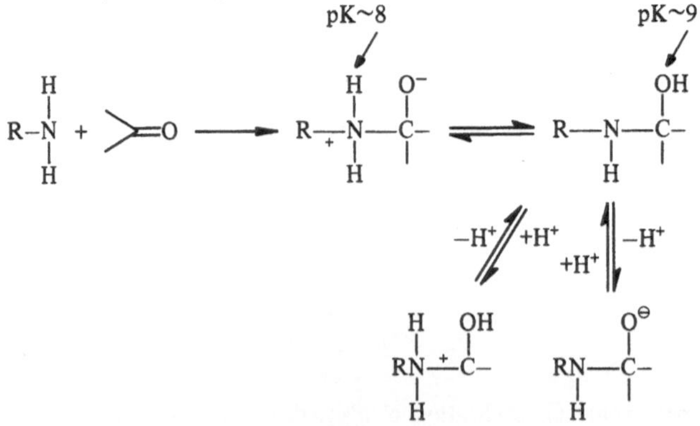

Scheme 2

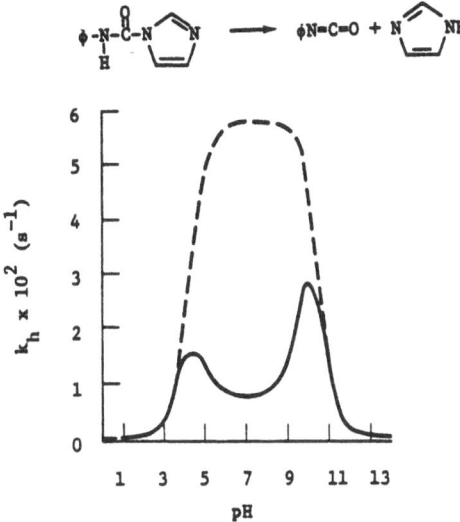

Fig. 8. k_h-pH profile for hydrolysis of 1-phenylcarbamoylimidazole at 30° (solid line). Dotted line represents profile if proton transfer is not rate determining [26]

Another complex pH profile is found in the hydrolysis of 1-phenyl carbamoylimidazole, *10* (Fig. 8) [26]

$$\phi-\overset{\text{O}}{\underset{\text{H}}{\overset{\|}{N}-C-N}}\diagdown$$

10

Below pH 4 the acid-promoted hydrolysis of the protonated species, (pK_A of imidazole ~ 4) is appreciably slower than hydrolysis of the neutral species, *10a*

10a   *10b*

Because of the negative charge on (*10b*) present at high pHs, the base catalyzed hydrolysis of anion, *10b*, is slower than hydrolysis of the neutral species. The curvature in the pH profile between pH 4–10 is explained by formation and reaction of the zwitterion *10c* whose chemistry is quite complex:

10   *10c*

$$\phi-N=C=O + N\diagdown NH \tag{29}$$

10c

   Hydrolysis of ethyl trifluorothiolacetate *11* [29] is exemplary of hydration effects in hydrolysis (Scheme 3)

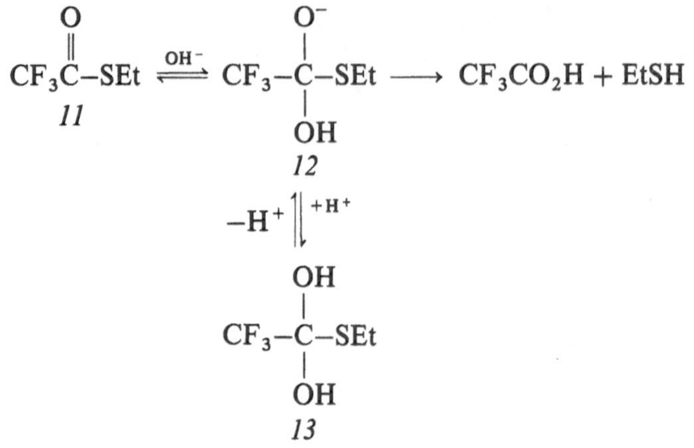

$$
\begin{array}{ccc}
\overset{\displaystyle O}{\underset{\displaystyle CF_3\overset{\|}{C}-SEt}{}} & \overset{OH^-}{\rightleftharpoons} & \overset{\displaystyle O^-}{\underset{\displaystyle CF_3-\overset{|}{\underset{|}{C}}-SEt}{}} \longrightarrow CF_3CO_2H + EtSH \\
\textit{11} & & \overset{\displaystyle}{OH} \\
& & \textit{12}
\end{array}
$$

$$-H^+ \Big\Updownarrow +H^+$$

$$
CF_3-\overset{\displaystyle OH}{\underset{\displaystyle OH}{\overset{|}{\underset{|}{C}}}}-SEt
$$

13

Scheme 3

   Hydrolysis occurs via the hydrate anion; at lower pH the rate decreases by formation of the stable diol.

   If both the neutral and ionized forms of a chemical (C and C', respectively) are present at significant concentrations at a certain pH and if C' hydrolyzes at a significantly different rate than C, the total rate of loss observed for a chemical with one ionizable group is

$$
\frac{-dC_T}{dt} = k_{hT}[C_T] = (k_A[H^+] + k_B[OH] + k_N)\,[C]
$$

$$
+ (k'_A[H^+] + k'_b[OH] + k'_N)\,[C'], \tag{30}
$$

where the prime denotes hydrolysis rate constants for the ionic species, the total chemical concentration $C_T = C + C'$ by mass balance equation and k_{hT} is the observed first-order rate constant for loss of total chemical. If C is an acid, then

$$
CH = C^- + H^+ \tag{31}
$$

and the equilibrium expression can be written[2]

$$
K_a = \frac{[C^-][H^+]}{[C]}. \tag{32}
$$

Equation (32) and the mass balance equation can then be combined to give

$$
C_T = C + \frac{K_a[C]}{[H^+]} = C\left[1 + \frac{K_a}{[H^+]}\right]. \tag{33}
$$

---

2   This expression also applies to the equilibrium $CH^+ = C + H^+$, where C is a base.

The fraction of chemical in the un-ionized form, f_c, is then

$$f_c = \frac{[C]}{[C_T]} = \frac{1}{1 + \dfrac{K_a}{[H^+]}} = \frac{[H^+]}{[H^+] + K_a}. \qquad (34)$$

The fraction in the ionized form, f_c', is then

$$f_c' = \frac{[C']}{[C_T]} = 1 - f_c = \frac{K_a}{[H^+] + K_a}. \qquad (35)$$

Combining equations (30), (33), and (34) gives

$$\begin{aligned} k_h^{obs} = &[k_A[H^+] + k_B[OH] + k_N]f_c \\ &+ [k_A'[H^+] + k_B'[OH]^- + k_N']f_c. \end{aligned} \qquad (36)$$

Equation (36) shows that, when $[H^+] \gg K_a$, hydrolysis of the unionized species controls k_h^{obs} ($f_c = 1$), when $K_a \gg [H^+]$ the ionic species is most important and when $pH = pK_a$ both species have equal concentrations.

Temperature Effects

Temperature affects hydrolysis reactions in the same way as other thermal reactions by increasing rates of hydrolysis with increasing temperature. Most investigators report temperature effects, which rarely span more than 60–70 °C, in the form of the Eyring or Arrhenius equations

$$\log k = -\Delta H^{\pm}/RT + \Delta S^{\pm}/R + C, \qquad (37)$$

$$\log k = A - E/RT, \qquad (38)$$

where A or $\Delta S^{\pm}$ are entropy terms and E or $\Delta H^{\pm}$ are temperature coefficients. Mabey and Mill [2] have included a range of values for E or $\Delta H^{\pm}$ in their compilation which can be summarized briefly as follows.

Neutral hydrolyses of alkyl halides have $\Delta H^{\pm}$ values which range between 90 and 100 kJ/mole. Acid catalysed hydrolysis of epoxides have $\Delta H^{\pm}$ values of 70–80 kJ/mole. Amides fall in this same range for both acid and base promoted reactions. Carbamates exhibit a wider range of $\Delta H^{\pm}$ values but the changes are not as clearly related to the values of k since $\Delta S^{\pm}$ also changes; values of 50 to 110 kJ/mole are reported [2].

As a practical matter, $\Delta H^{\pm}$ of 50 kJ/mole corresponds to doubling of the rate for a 10 °C change near 25 °C while 100 kJ/mole corresponds to a four fold increase over the same temperature span. Since many values of $\Delta H^{\pm}$ are not known a safe assumption for carbonyl reactions would be that the rate changes by a factor of 2 to 3 for each ten degree change in temperature.

Another important effect of temperature in hydrolysis reactions is on pH. Water is more dissociated at higher temperature and the value of K_w increases according to the relation

$$\log K_w = -6013.79/T - 23.652 \log T + 64.7013. \qquad (39)$$

The other effect of temperature is on pH of buffer systems. If pH is adjusted to the same value at different temperatures, the concentration of OH^- ion will be different because of the change in K_w:

$$pW = pH + pOH + C. \tag{40}$$

Properties of Natural Waters

The usefulness of hydrolysis kinetic constants in water for estimating environmental fate depends also on reliable information on the properties of natural fresh and marine surface waters and groundwater. The primary environmental factors which control the value of kinetic constants in the environment are pH and temperature. Secondary factors such as ionic strength, metal cation and anion composition and content and dissolved or suspended organic matter also may influence hydrolysis of some chemicals. This section reviews the average or range of properties of fresh and marine waters. This information coupled with our discussion of pH and temperature in the previous section and of ionic effects, metal ion catalysis and sorption in the following sections will assist the user of kinetic information in making choices and calculations appropriate to the surface or ground water of interest.

Ionic Composition of Fresh and Ground Water

Trace metal ions are difficult to measure accurately in natural waters. Only in the past 10 years have techniques become available with which to sample and analyze for metal ions with any confidence in the results [30]. More abundant cations and anions are widely measured as criteria of water quality and in the US are a matter of record in the Geological Survey [31]. Table 2 summarizes the best recent estimates for a variety of cations and anions with the *caveat* that data on trace metal ions older than 10 years may be suspect. No information is available on speciation of metal ions in these water bodies but models of systems containing varying proportions of metal ions and fulvic acids suggest that both in fresh and marine waters, high proportion of trace metals exist as aquo or chloro complexes because more abundant divalent cations (Ca^{2+} or Mg^{2+}) compete for humic acid sites and Cl^- ion competitively complexes trace metals in seawater [32].

Ionic Composition of Seawater

The major ionic composition of seawater is relatively constant from place to place and accordingly the pH is close to 8.3 almost everywhere. Only alkali and alkaline earth cations are abundant (Table 2). Major anions in order of abundance (molar) in seawater are shown below [33]

Cl^-	0.53
SO_4^{2-}	0.028
HCO_3^-	0.0024
Br^-	8.2(−4)
F^-	6.8(−5)
$H_2BO_3^-$	4.0(−4)

Table 2. Metal ion abundance in waters in molar concentration[a]

	Na	K	Ag	Ca	Mg	Al	Zn	Hg	Fe	Mn	Cu	Ni
Rain water	>1.7(−5)	>7.7(−7)	—	2.5(−6) 2.5(−4)	>4.2(−6)	—	—	—	—	—	—	—
Rivers	1.1(−3)	5.1(−5)	—	9(−4)	3.8(−4)	—	—	—	1.1(−6)	—	—	—
Lake chlorides	7.6(−4)	3.5(−5)	—	2.5(−4)	2.7(−4)	—	—	—	—	—	—	—
Lake sulfate	4.9(−4)	3.5(−5)	—	1.5(−4)	4.7(−4)	—	—	—	—	—	—	—
Lake carbonate	1.1(−3)	1.7(−4)	—	1.5(−4)	1.5(−4)	—	—	—	—	—	—	—
Lake surface	3.1(−2)	9.7(−4)	—	1.4(−3)	4.2(−3)	—	1.5(−7)	—	1.4(−6) 7.1(−7)	6.4(−7)	1.6(−7)	—
Subsurface lake	2.8(−3)	8.9(−5)	—	2.0(−3)	1.1(−3)	—	—	—	5.4(−6)	—	—	—
Maine lakes[b]			2.6(−7)				3.8(−3) 5.2(−7)	3.3(−10)		3.6(−10) 1.6(−6)	1.0(−9) 2.2(−6)	1.7(−10) 1.2(−7)
Fresh waters[c]						1.4(−5) 3.3(−3)	7.7(−9) 1.5(−7)					
Ocean water[c−e]	4.8(−1)	1.0(−2)	2.5(−11) 2.7(−9)	1.0(−2)	5.4(−2)	2.0(−8) 2.6(−6)	6(−9) 4.6(−8)	5(−12) 3.3(−10)	5(−9) 8.9(−8)	5.1(−10) 9.1(−8)	4.7(−9) 2.7(−8)	3.4(−9) 2.5(−8)

a Blessing and Mill [34]; values in parentheses are powers of ten
b Horne [35]
c Florence and Bately [36]
d Stumm and Morgan [37]
e Bruland [30]

Ionic Strength and Salt Effects

Solutions with high concentrations of ionic species can increase rates of hydrolysis if the reaction involves bond breaking with charge formation at the reaction center as the leaving group X departs [3]:

$$-\overset{|}{\underset{|}{C}}-X \longrightarrow -\overset{|}{\underset{|}{C}}{}^{\delta+}----X^{\delta-}, \tag{41}$$

$$-\overset{|}{\underset{|}{C}}{}^{\delta+}-----X^{\delta-} + H_2O \longrightarrow -\overset{|}{\underset{|}{C}}-OH + HX. \tag{42}$$

The ionic strength of a solution, μ, is defined by Eq. (3)

$$\mu = (\Sigma C_i Z_i^2)/2, \tag{43}$$

where C_i is the concentration of the ith ion and Z_i is its charge. The effect of ionic strength on the hydrolysis rate is attributed to the ionizing power of the solvent either to stabilize or destabilize a solvated reaction intermediate relative to the solvated starting reactants (1). Reactions that involve no significant charge separation or generation in going from uncharged (or neutral) reactants to intermediate are generally little affected by the ionic strength of reaction media (see below). The true effects of ionic strength are sometimes difficult to measure because ionic strength also is important in determining pH. Thus changes in hydrolysis rates at different ionic strengths may be due to either or both pH and ionic strength effects. Ionic strength also may increase or decrease rates of non-ionic reactions, however, the direction and magnitude of the effect is not generally predictable other than that, in general, the effect is usually less than a factor of 2. Perdue and Wolfe [38] have concluded that simple ionic strength effects are not important in natural waters for most chemicals.

Frost and Pearson [39] cite an example of an ionic strength effect on the acid-catalyzed hydrolysis of γ-butyrolactone, a reaction involving a neutral molecule and an ion

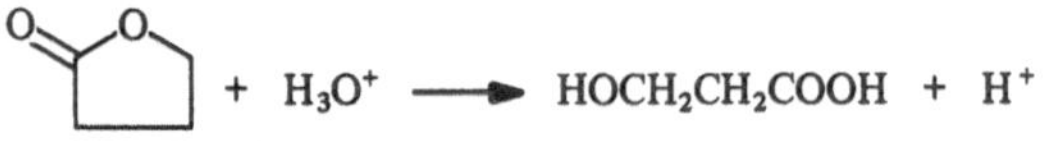

$+ \ H_3O^+ \ \longrightarrow \ HOCH_2CH_2COOH \ + \ H^+$ $\tag{44}$

Salt effects of NaCl and NaClO$_4$ are in opposite directions, and the differences are attributed to differing effects on the activity coefficient of the lactone. Both SO_4^{2-} and Br$^-$ should also exhibit positive salt effects in the hydrolysis of γ-butyrolactone, by analogy to the effect of Cl$^-$. However, in sea water, where μ is less than 0.4, these salt effects are insignificant.

Mass Law Effects

Hydrolysis reactions conducted in solutions that already contain high concentrations of the leaving group X can result in reformation of reactant in an exchange process

$$R-X + X^* \longrightarrow R-X^* + X, \tag{45}$$

where the * labels the incoming group and X is the displaced group. The overall effect of [X*] on the hydrolysis rate constant depends on the extent of R–X bond breaking before X* interacts with R to form R–X* (which of course is analytically identical with R–X). If the reaction occurs by simple displacement of X by X*, no net change in bonding has occurred (e.g., R–X bond breaking is compensated for by R–X* bond making). There is no effect on the overall loss rate of reactant because the loss of R–X is offset by formation of R–X*

$$-\frac{d(R-X)}{dt} = k_h[C] + \frac{d(RX^*)}{dt} - \frac{d(RX)}{dt}. \tag{46}$$

However, if RX solvolyzes via a S_N1 process to form an intermediate ion-pair, the presence of X in solution can slow the overall loss of R–X. This effect, known as the mass law effect, depends on the total amount of X present and the stability of R^+. The reaction sequence is:

$$RX \xrightarrow{k_1} \overline{R^+X^-} \quad \text{(ion pair)}, \tag{47}$$

$$X^* + \overline{R^+X^-} \xrightarrow{k_2} RX^* + X, \tag{48}$$

$$H_2O + \overline{R^+X^-} \xrightarrow{k_3} ROH + X. \tag{49}$$

The rate of loss of RX is then given by Eqs. (50) and (51),

$$-\frac{d(RX)}{dt} = \frac{k_1 k_3 [RX]}{k_2[X]^- + k_3} \tag{50}$$

$$= \frac{k_1[RX]}{\dfrac{k_2}{k_3}[X] + 1}. \tag{51}$$

These equations show that when $k_2[X] \ll k_3$, hydrolysis with water follows simple first order kinetics. However, when $k_2[X] \gg k_3$, the rate of loss of RX is inversely dependent on $[X]^-$. The ratio k_2/k_3 is called α.

It is the most stable carbonium ion-forming halides which have the greatest sensitivity to added halide ions. Triphenylmethyl chloride has an α value of approximately 400; evidently this most reactive alkyl halide exhibits the greatest discrimination between chloride ion and water in the ion pair. Similarly, much less reactive t-butyl bromide exhibits almost no discrimination. In seawater, where chloride ion concentration is approximately 0.3 M, triphenylmethyl chloride would hydrolyze approximately 120 times slower than in pure water. Table 3 summarizes values of α estimated by Swain and Scott [19].

For benzhydryl chloride in seawater, where ($[Cl^-] = 0.3$ M and $\alpha = 10$ (ionic strength effects on α is ignored in these calculations), its rate loss of is about one-quarter of that in pure water. However, for t-butyl chloride, where $\alpha = 0.4$ the hydrolysis rate in sea water would be more than 90% of the rate in pure water.

Because, benzhydryl and t-butyl chlorides represent extremely reactive halides ($t_{1/2} = 1–3$ h) and yet show only small influences of mass law effects, we conclude that generally the mass law effect is not important under environmental conditions.

Table 3. Values of mass-law constant, α, for several alkyl halides in 85% acetone-H_2O at 25°[a]

Alkyl halide	α
Triphenylmethyl chloride	~ 400
p-Methylbenzyhydryl chloride	28–35
Benzhydryl chloride	10–16
Benzhydryl chloride	50–70
t-Butyl bromide	0–2[a]

[a] Swain and Scott [19]

An experimental test of this conclusion is found in the recent study of Mabey et al. [40] on the hydrolysis of benzyl chloride in pure water and in artificial seawater. Benzyl chloride has a hydrolysis half life of about 15 h at pH 7 and 25 °C (2). The rate is pH independent from pH 0–11 and the kinetics are well-characterized by Robertson and Scott [41]. Despite the pH independence of the rate constant and the relatively high sensitivity of the rate constant to substituents in the aromatic ring [42] the activation parameters of Robertson and Scott [41] suggest that hydrolysis of chloride occurs via a S_N2 transition state involving water. The rate constant measured at pH 7 to 8 in pure water and seawater differed by only 3%, just barely outside the error limits of the measurements. Clearly, there is no significant ionic strength nor mass law effect on this hydrolysis reaction. Replacement of chloride ion by perchlorate ion in seawater gave only a 10% reduction in rate constant; similarly addition of KCl to pure water or 0.05 M phosphate buffer gave no significant change in rate constant.

Similar studies of the hydrolysis of 1,3-dichloropropene at 35° in pure water, artificial seawater and artificial seawater with perchlorate ion instead of chloride ion, all gave nearly the same rate constant within 10% [40].

Metal Ion-Catalyzed Hydrolysis

Metal ions such as Ca^{2+}, Cu^{2+}, Co^{2+}, and Zn^{2+} can act as catalysts for hydrolysis reactions involving many classes of hydrolyzable chemicals [42]. The widespread occurrence of many of these same metal ions in fresh and marine waters suggests that metal-ion catalysis could be important under some conditions in the environment with certain structural classes.

Organic-Metal Complexes

Martell [43] notes that metal ions are strongly bound by amines and carboxylate, enolate, phenolates, mercaptide, and phosphorous anions; ligands which may weakly coordinate with some metal ions include carbonyls, ethers, esters, amides, and thioamides. Weaker ligands become more effective when stronger ligands also are present on the same structure, e.g., multidentate ligands such as EDTA. In aqueous systems, metal ions are coordinated by water molecules as aquo complexes, $M(H_2O)^{n+}$; the actual number of water ligands varies with the metal and other available ligands (L) [e.g., $ML_3(H_2O)_{x-3}$]. For each metal-ligand aquo complex, a specific formation rate constant, k_f, and stability constant K_{ML} apply,

$$M + L \xrightleftharpoons[k_r]{k_f} ML \tag{52}$$

$$K_{ML} = \frac{[ML]}{[M][L]} = \frac{k_f}{k_r}, \tag{53}$$

$$M(H_2O)_x + L \longrightarrow ML(H_2O)_{x-1} + H_2O, \tag{54}$$

and each complex may have a different ability to catalyze hydrolysis. In natural waters, humics provide ligands for many metal ions binding them through carboxylate, phenolate, and other groups common to humic and fulvic acids; as a result, metal-ions in natural waters may exhibit significantly different catalytic activity than those in pure water. In seawater chloride ion is a major complexing species.

Several literature citations suggest that metal-ion promoted hydrolysis can be important in the environment but these studies involved metal-ion concentrations much higher than those found in aquatic systems [43, 44]. Furthermore, several papers have reported that in aquatic environments some metals exist in the colloidal form rather than in true solution [45].

Surprisingly little is known about the concentration, distribution, and speciation of metal in major rivers and lakes. Some literature data on metal-ion concentrations are summarized earlier. These data represent total metal-ion concentrations without regard to speciation of the metal which varies widely from place to place.

In general, freshwater systems contain substantial concentrations of alkali and alkaline earth cations ($> 1 \times 10^{-4}$ M), moderate concentrations of Fe ($> 1 \times 10^{-6}$ M), but largely unknown amounts of Al, Hg, Zn, Ni, and Mn. In contrast, marine waters have been well characterized for metal ions, and most are at very low concentrations ($< 1 \times 10^{-7}$ M), except Na, K, Ca, Mg, and Al. Zinc is the one exception and seems to be as abundant in marine waters as in fresh water.

Mechanism of Metal Ion Catalysis

Although an oversimplification, it is convenient to categorize metal catalyzed hydrolysis mechanisms either as hydroxide ion carrier reactions in which the metal-aquo complex acts as a HO^- transfer agent at pH's, where free OH^- is not important or as complexation reactions, where the metal-ion complexes with the organic so as to facilitate hydrolysis by HO^- or H_2O

$$M(H_2O)^{(n+1)+} \longrightarrow M(H_2O)^n + (OH) + H^+, \tag{55}$$

$$SX + M(H_2O)^{n+}(OH) \longrightarrow SOH + M(H_2O)^{n+}X, \tag{56}$$

$$M + SX \rightleftharpoons MSX \xrightarrow{H_2O, OH^-} S{-}OH + M + X^-. \tag{57}$$

The HO^- carrier process has been investigated by Buckingham and coworkers using Cu, Co, and Zn [46, 47]. The concentration of the hydroxide species $(M(OH)^{n+})$ at a given pH is determined by K_A of reaction (4)

$$(MOH)^{n+} = \frac{K_A[M_T]}{K_A + [H^+]} \tag{58}$$

where M_T is total metal ion concentration and k_h follows the relation

$$k_h = k_A[H^+] + k_N[H_2O] + k_B K_A[M_T]/(K_A + [H^+]). \tag{59}$$

We evaluated the importance of the HO^- carrier process in the environment using data for the Zn-catalyzed reaction of 2,4-dinitrophenyl acetate at pH 7 [46] and Zn concentrations of 1×10^{-6} M. At pH 7 the reaction is controlled entirely by k_N; at pH 9 $k_B[OH]$ contributes 5% of the reaction and Zn contributes less than 1% at both pHs. Clearly, the Zn^{2+}-catalyzed process will be unimportant at the low concentration of metals commonly found in aquatic environments. The metal-ion-catalyzed hydrolysis of parathion 14, by Ca^{2+} and Cu^{2+} reported by Ketalaar et

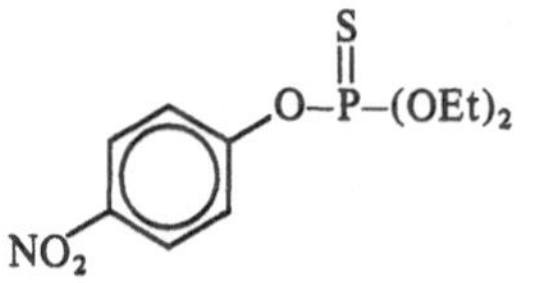

14

al. [48] may be an example of a complexation process; however, insufficient kinetic data are available to rule out the HO^- carrier process. A possible example of direct catalysis by Mg^{2+} and Zn^{2+} in hydrolysis of the phosphate esters 15, 16, and 17 is reported by Steffans et al. [49]

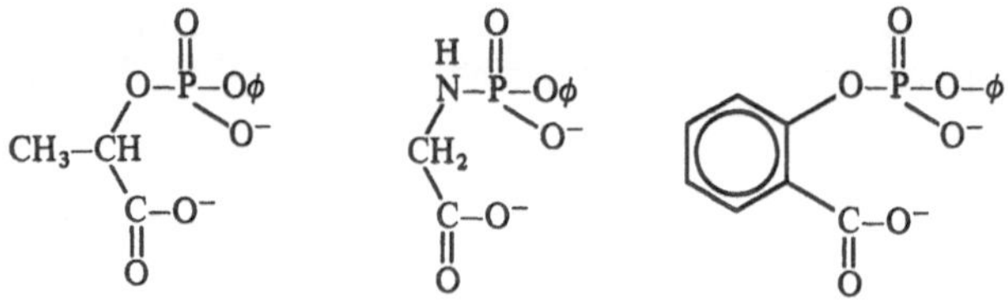

15   16   17

Because both the catalyzed and the uncatalyzed hydrolyses occur via a cyclic acyl phosphate intermediate, the authors argue that the metal ion stabilizes a pentacovalent intermediate 18

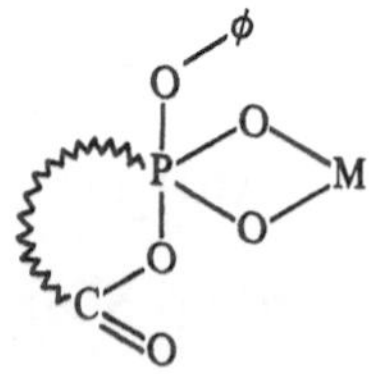

18

However, we calculate that with Zn or Mg concentrations of 1×10^{-6} M, the contributions of the metal-catalyzed processes to the total calculated rate constants are only 0.5–2% for Zn^{2+} and $<0.06\%$ for Mg^{2+}.

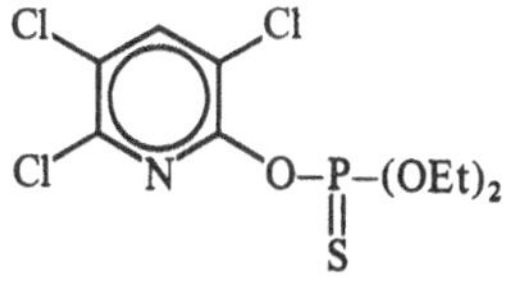

19

Several groups have reported that the pesticide chlorpyrifos, *19*, or its methyl ester, is catalyzed by cupric ion [50–52]. Blanchet and St. George [50] reported that, with Cu^{2+} concentrations above 11 mM and chlorpyrifos concentrations of 2–3 μM, the hydrolysis rate was independent of Cu^{2+}. At lower Cu^{2+} concentrations, the hydrolysis was first order in metal ion and in pesticide.

$$-\frac{dP}{dt} = k_2[P][Cu^{2+}] = k_h[P], \qquad (60)$$

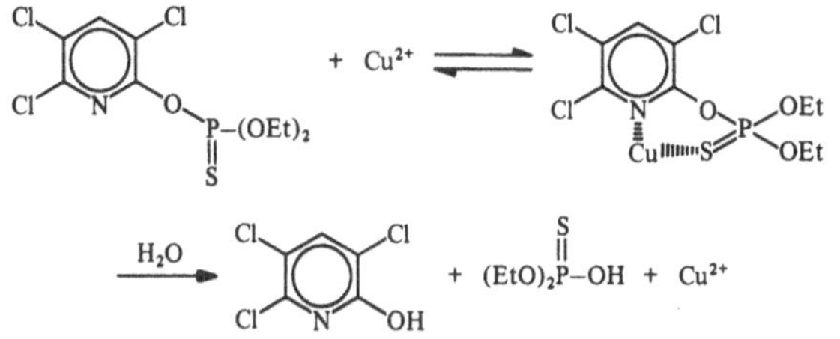

$$(61)$$

Increasing the pH from 4.4 to 5.8 increased the hydrolysis rate, but Blanchet and St. George reported no specific kinetic order for [OH$^-$].

Meikle and Youngson [51] also studied the hydrolysis of chlorpyrifos at several pHs with and without Cu^{2+} present and report a 4 to 10 fold increase in the rate constant with metal ion. The data of Blanchet and St. George [50] are in fair agreement with that of Meikle and Youngson [51]. However, using 10^{-6} M Cu^{2+} gives a Cu^{2+} catalyzed rate constant ($k_h = 0.0015$ d^{-1}) which is not competitive with the uncatalyzed process ($k_h = 0.011$ d^{-1}). The metal-catalyzed hydrolysis

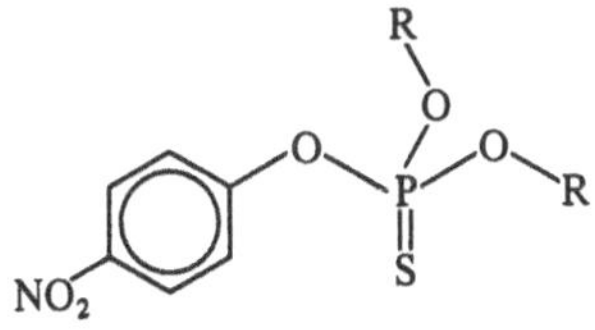

20

of parathion (O,O-dimethyl-p-nitrophenyl thiophosphate *20*, PT, R = Et) is reported by Ketalaar et al. [48]. He claimed that the calcium- and copper-catalyzed reactions of parathion were significant even at metal ion concentration of 10^{-8} M. A complex in which Cu binds preferentially at the sulfur and alkoxy oxygen atoms (*21*) may account for the catalysis.

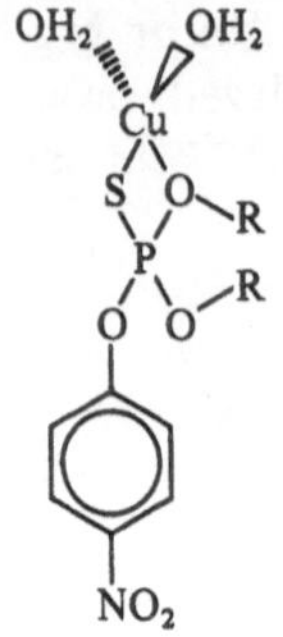

21

Methyl parathion (MPT, R=Me) has an uncatalyzed, pH-dependent hydroly-sis in which the hydrolysis rate is constant from pH 2 to 8 [53]. Below pH 8, hydrolytic C—O cleavage is favored, yielding greater than 95% of monomethyl-p-nitrophenyl thiophosphoric acid and methanol. Above pH 8, P—O cleavage by OH^- is favored and p-nitrophenol and dimethylthiophosphoric acid become the dominant products. These results are consistent with the concept that the phosphoryl-P is a hard acid center preferentially attacked by a hard hydroxyl ion nucleophile, whereas tetrahedral carbon is a soft acid center attacked easier by the softer nucleophile water.

Mabey et al. [40] measured kinetics of hydrolysis of MPT with 1×10^{-4} M MPT and with 0 to 1×10^{-4} M Cu^{2+} and without buffers. MPT hydrolysis shows increasing rates with increasing Cu^{2+} concentrations but added humic acid inhibits the reaction with only 1×10^{-5} M Cu^{2+}. Apparently Cu^{2+} promotes the cleavage of the P—O bond even at low pH; mostly nitrophenol is formed with Cu^{2+} present.

Mabey et al. [40] also examined the possible catalytic effects of 1×10^{-4} M Cu^{2+}, Ca^{2+}, and Zn^{2+} on hydrolysis of fourteen esters, amides or nitriles which also had strong metal-binding ligands in their structures, including pyridine-N, carboxylate and phenolate. Of the thirty-three rate constant ratios of k(metal)/k(control) measured at 70 °C, only one ratio was larger than ten: 2-ethyl picolinate (ED) with Cu^{2+}. No significant effects were observed with Ca^{2+} or Zn^{2+}. The catalytic rate constant for ED and Cu^{2+} is large but 10 ppm added humic acid largely inhibits the reaction. These results strongly suggest that environmental metal ions catalyze hydrolysis only of unusual chemical structures; most hydrolyzable molecules will not exhibit metal-ion catalysis in natural waters.

Effects of Sorption and Surfaces

Neutral organic chemicals sorb to soil or sediment by partitioning into the organic carbon fraction; good correlations of the partitioning constant for sediment (K_{oc}) are found with the octanol-water constant (K_{ow}) or with water solubility. Karickhoff [54] has reviewed sediment sorption and kinetics recently and Karickhoff et al. [56], Yalkowski and Volvani [57], and Chiou et al. [55] have developed several regression equations for calculations of k_{oc} based on K_{ow} or water solubility.

Hydrolysis on Sediment

The simple assumption that an organic chemical sorbed to sediment is sequestered and unavailable for hydrolysis appears to be only partly true. In a slurry of sediment and water the equilibrium concentration of chemical in the water, C_W, is

$$C_W = \frac{C_T}{1 + \varrho K_p}, \tag{62}$$

where C_T is total chemical concentration, ϱ is the sediment to water ratio and K_p is the ratio C_s/C_W related to K_{OC} by K_p O.C. If hydrolysis occurs only in the water phase (rate constant k_W)

$$k_{Obs} = k_W/(1 + \varrho K_p). \tag{63}$$

Maclady and Wolfe [58] showed that several organophosphorothioates including chloropyrifos, diazinon and ronnel, hydrolyze by a neutral process about as rapidly in the sediment as in the water phase.

Similar results were found for water substitution with sorbed n-alkyl bromides (S_N2), with benzyl chloride (S_N1) and with hexachlorocyclopentadiene (S_N2'). However, alkaline hydrolysis of a phosphorothioate ester (chloropyrifos) and alkyl esters of 2,4-dichlorophenoxyactic acid (2,4-D) gave no reaction in the sorbed state. Acid hydrolysis of the same 2,4-D ester apparently was faster in a slurry of sediment at low pH (no detailed results were given). Alkaline sediments are negatively charged (from carboxylate and cationhydrate anions) which may account for inhibition of the base promoted process. But why neutral and acid processes proceed at normal or accelerated rates is unknown and serves as a cautionary note in evaluating effects of sediment sorption on hydrolysis.

Hydrolysis on Soils

Hydrolysis in soils has been investigated by Sethuan and MacRae [59], Mingelgrin et al. [60], and El-Amamy and Mill [61]. The situation is complicated by a pronounced increase in surface acidity at low water content in clay soils. Enhanced acidity occurs between 3 and 20% water owing to strong polarization of the hydrated cations [60, 61]. Acid catalyzed hydrolysis of phosphorus esters and epoxides were observed on montmorillonite and kaolinite with rate constants corresponding to a pH 2 units lower than the bulk medium pH [60, 61]. Soil with moisture constants higher than 25% (based on soil dried at some fixed temperature) show no enhanced rates. None of the compounds examined sorb strongly to clay surfaces (non-ionic) or to the organic fraction (low log Ps). In real soils where some chemicals may partition to organic carbon and some to clay surface, depending on the soil and the chemical structure, prediction of how rapidly hydrolysis will occur or where is an uncertain process.

Hydrolysis on Dissolved Humics

Dissolved organic material (humic acid) present in natural waters contains a wide variety of organic structures, including phenols and aromatic acids, quinones, lignin, carbohydrates, alcohols, amino acids and fatty acids [62]. These structures contribute to the polyelectrolyte nature of the dissolved organic matter found in soils and aquatic systems. Two classes of polyelectrolytes, called humic (HA) and fulvic acids, (FA) are normally found at much higher concentrations (several ppm) than other polyelectrolytes. In fresh waters these HA and FAs are present in mass concentrations as high 20 ppm in very entrophic waters. Carboxylates in both materials are substantially ionized at pH > 5.

Humic and fulvic acids may physically adsorb other chemicals by either hydrogen or van der Waals bonding [63]. Some hydrophobic chemicals may be associated with dissolved humic substances in solution. This association has been used to explain chemical concentrations that exceed solubility limits in pure water [64], decreased availability of chemicals as measured by toxicity tests, and the failure of some medium molecular weight chemicals such as DDT to pass through membranes [64].

Landrum et al. [65] showed that partitioning of dissolved compounds to dissolved humic acids can be estimated using reverse phase HPLC much in the same way as K_{ow} is evaluated which suggests that a partitioning process is responsible for the phenomena. No evidence has been found that "soluble partitioning" affects hydrolysis rates; however, very few experiments designed to test for effects are reported.

Laboratory Measurements of Hydrolysis Rate Constants

Several documents have been prepared in the past three years which describe detailed procedures for measurements of hydrolysis rate constants for organic compounds in water in the laboratory under conditions where the rate data can be used to estimate rate constants at any pH of interest [18, 40, 66]. The basis for this series of measurements rests on the Eqs. (1) through (4) in the Kinetic Relations section. Two kinds of measurements are proposed: a screening test which has the objective of identifying chemicals with half lives of less than one year on hydrolysis at 25 °C and pH 5–9 and a detailed test which provides a procedure for accurate measurements of rates of hydrolysis at any pH and temperature commonly encountered in natural waters-usually pH 5–9 and 5–30 °C. In this test, dependences of k_h on both pH and temperature are developed which than are used to estimate values of k_A, k_B, and k_n as well as the temperature dependence of the individual reactions.

Hydrolysis Test Protocols

The screening test involves simple measurements of rates of loss of chemical in water solution below the solubility limits of the chemical at 25 °C in sterile dark solutions. pH is maintained by the use of buffer solutions at pH 3, 7, and 11. Samples are withdrawn at intervals of 0, 44, and 88 h. A simple first order plot of log concentration versus time at each pH provides a crude but satisfactory basis for estimating the observed first-order rate constant k_h.

With this information (coupled with satisfactory control solutions kept in the dark at 5 °C at pH 7) the rate constants may be interpreted in the following way: if more than half the initial concentration of chemical is hydrolyzed at pH 3 or 11 or both in 88 h, the chemical will have a half life of less than 1 year at pH 5 or 9 or both. The basis for this screening study is that many chemicals will hydrolyze at pH 3 or 11 about 100 times faster than at 5 or 9; thus the measured half life of less than 88 h at 3 or 11 is equivalent to half lives of less than 8,800 h or 1 year at pH 5 or 9. Similarly if more than 75% of the chemical hydrolyzes after 2 h the half life is less than one hour. Generally, detailed studies are required only if the half life is found to be less than a year at pH 5–9.

The purpose of the detailed test is to provide accurate data on pH profile and temperature dependence. The procedure differs from the screening test in several important respects: multiple data points are taken as a function of time with not less than 6 independent data points corresponding to 20 to 70% hydrolysis. If hydrolysis is too low or too fast to measure in 1 h to 2–3 weeks, a lower or higher temperature should used, in which case data analysis becomes more complex in order to ensure an accurate extrapolation of measured values at the lower or higher temperature to 25 °C. Temperature effects were discussed in detail in the second section; it is sufficient here to note only that rate constant measurements at temperatures significantly greater than 15° above or below 25° are better extrapolated back to 25° first by analyzing the observed rate constants in terms of the contributions from the acid, base and neutral processes, (k_A, k_N, and k_B) estimating the Arrhenius parameters for each process independently, extrapolating these individual rate constants back to 25° and then recombining them to give the observed rate constant (k_h) at 25°.

Temperature Dependence

The temperature and pH dependence of hydrolysis of ethyl acetate are shown in Table 4. The results suggest that for rate constants measured at two or three temperatures between 25 and 75 °C, detailed analysis and extrapolation of individual process constants as opposed to extrapolation of the observed rate constant lead to a difference in rate constant of only 20% over this temperature range. Although this is a small difference the value of k_h at 25 °C obtained by extrapolating k_h values at higher temperature cannot be used to calculate k_h at pH values other than those for which the temperature measurements have been made. The more exact method of estimating the temperature dependence of the process rate constant will give k_h at any pH and is preferred.

The hydrolysis of ethyl acetate serves as an example of the method. The rate constants, k_h, for hydrolysis of ethyl acetate at three pHs for reactions at 25°, 55°, 75°, and 95 °C are summarized in Table 4. For each hydrolysis experiment, three replicates were analyzed for each time point, and the average of these analyses was used in the regression analysis to obtain the first-order rate constants k_h (in s^{-1}). Hydrolysis at pH 11 and 55 °C or 75 °C was too fast to measure. Therefore, high pH rate constant measurements at 55 °C and 75 °C were measured at pH 10 and 9.

Table 4. k_h Values for hydrolysis of ethyl acetate (EA)

Rate constant	pH	10^3 (EA), M	Number of time points	$k_h,^a s^{-1}$	% conversion	r^2	Total elapsed time
25°C							
$k_{h(3)}$	3.00	0.998	10	$(1.73 \pm 0.15) \times 10^{-7}$	38	0.97	36 d
$k_{h(7)}$	7.00	1.03	2	6.98×10^{-8}	25	–	88.8 d
	7.00	1.01	2	4.12×10^{-8}	39	–	138.9 d
$k_{h(11)}$	11.00	1.04	7	$(1.08 \times 0.01) \times 10^{-4}$	78	0.999	3.8 h
55°C							
$k_{h(3)}$	2.84	1.01	7	$(1.92 \pm 0.02) \times 10^{-6}$	85	0.999	11.8 d
$k_{h(7)}$	7.00	1.01	6	$(7.24 \pm 0.21) \times 10^{-7}$	68	0.996	16.8 d
$k_{h(10)}$	9.86	1.01	7	$(3.12 \pm 0.16) \times 10^{-4}$	87	0.985	1.74 h
75°C							
$k_{h(3)}$	3.00	1.01	7	$(9.15 \pm 0.40) \times 10^{-6}$	82	0.988	48.3 h
$k_{h(7)}$	7.00	1.01	6	$(5.66 \pm 0.18) \times 10^{-6}$	67	0.995	52.1 h
$k_{h(9)}$	8.69	1.01	7	$(2.31 \pm 0.06) \times 10^{-4}$	90	0.996	2.85 h
95°C							
$k_{h(3)}$	3.15	1.01	6	$(2.29 \pm 0.03) \times 10^{-5}$	84	0.993	22.1 h
$k_{h(7)}$	7.00	1.01	5	$(3.48 \pm 0.11) \times 10^{-5}$	31	0.997	12.9 h

[a] Rate constant not corrected for pH change due to inadequate buffering

In direct extrapolation of k_h the data are fitted at four temperatures for $k_{h(3)}$ and for $k_{h(7)}$ to the Arrhenius equation in the form (procedure 4)

$$\log k_h = \log A_h - \frac{E_h}{2.3 R} (T)^{-1} . \tag{64}$$

The values for log A and E_h/R are listed below

	LogA	E_h/R	r^2
$k_{h(3)}$	4.62 ± 0.43	$7,807 \pm 327$	0.997
$k_{h(7)}$	7.94 ± 0.60	$10,548 \pm 458$	0.996

The fit of the data is quite good considering that the relative contributions of the acid and neutral processes may change with temperature at pH 3 and that the relative contributions of the neutral and base processes may change with temperature at pH 7. The good fit of the pH 7 data is surprising because the autoprotolysis equilibrium constant K_w increases with temperature and with it the concentration of OH^-. Thus at pH 7 and 75°C, $[HO^-]$ is 2.04×10^{-6} M, and the contribution of the base-catalyzed process should increase with temperature

$$pK_w = pH + pOH . \tag{65}$$

Values of $\log K_w$ at the temperatures of these experiments were calculated from the equation given in Harned and Owen [67]

$$\log K_w = \frac{-6013.79}{T} - 23.652 \log T + 64.7013 . \tag{66}$$

In the three-step procedure, values of k_A, k_B, and k_N at 55 °C and 75 °C were calculated from values of k_h by solving the simultaneous equations with k_h values at pHs 3, 7 and 9 or 10 and estimating E_a and A from Eq. (64). The Arrhenius parameters for k_A, k_B, and k_N calculated from these data are shown below. Units for A are the same as for k

	E_a (kcal/mole)	A
$k_A(M^{-1}s^{-1})$	14.30	8.318×10^6
$k_N(s^{-1})$	9.04	1.59×10^{-1}
$k_B(M^{-1}s^{-1})$	13.58	8.73×10^7

From these Arrhenius expressions for k_A, k_B, and k_N, rate constants $k_{h(3)}$, $k_{h(7)}$, and $k_{h(11)}$ were calculated at 25 °C. These rate constants are listed below along with similar values that were directly measured or calculated via procedure 4

Method	$k_{h(3)} \times 10^7$	$k_{h(7)} \times 10^8$	$k_{h(11)} \times 10^4$
Direct measure	1.76	4.12	1.26
Direct extrapolation (via E_h)	1.25	1.97	1.65
3-Step extrapolation (via E_A, k_B or k_N)	3.12	4.74	0.97

Since rate constants $k_{h(11)}$ were not measured at 55 °C and 75 °C, these values used in direct extrapolation procedure for $k_{h(11)}$ were calculated by multiplying the $k_{h(10)}$ and $k_{h(9)}$ values by 10 and 100, respectively, then calculating and using the Arrhenius expression to obtain $k_{h(11)}$ at 25 °C.

The relative contributions of the k_A, k_B, and k_N terms to k_h as a function of pH and temperature were calculated from $k_A[H^+]/k_h$, k_N/k_h, etc. The k_B term dominates at all temperatures at pH 9 and 11. At pH 3, the k_N terms increase in importance in going from 75 °C to 25 °C, but k_h is still dominated by the k_A term. However, the base-promoted process becomes a more dominant contributor to $k_{h(7)}$ at higher temperature, and thereby demonstrates the problem with the direct extrapolation procedure at pH 7.

The important feature of these procedures is their emphasis on conditions most likely to yield reliable rate constants for hydrolysis in water or water containing small amounts of acetonitrile (used to ensure solubility). This is achieved by the use of buffer solutions to maintain pH and controls to minimize the likelihood of adventitous processes which would confound interpretation of the data. In addition, specific reporting procedures are recommended to ensure the measured data are provided as sufficient background detail to give other investigators confidence in measurements. Hydrolyses data which lack sufficient experimental detail by which to judge whether competing processes such as volatilization or biodegradation are not used.

Buffer solutions used to control pH in these reactions are listed in Table 5. The reader should note that buffers should not be relied on to provide exact pH values even with close control of concentrations and temperature. pH values should always be measured after preparation of the buffer using a good pH-meter calibrated with standard buffers near pH 4, 7 and 9.

Table 5. Standard buffer concentrations used in hydrolysis experiments

pH 3.00 SBC[a]	500.0 mL of 0.1 M potassium hydrogen phthalate ($KCH_6H_4O_4$) and 223.0 mL of 0.1 N HCl, diluted to 1.000 L with water[b]
pH 7.00 SBC[a]	500.0 mL of 0.1 M potassium dihydrogen phosphate (KH_2PO_4) and 291.0 mL of 0.1 N NaOH, diluted to 1.000 L with water[b]
pH 11.00 SBC[a]	500.0 mL of 0.05 sodium bicarbonate ($NaHCO_3$) and 227.0 mL of 0.1 N NaOH, diluted to 1.000 L with water[b]
pH 10.00 SBC[a,c]	500.0 mL of 0.05 M sodium bicarbonate ($NaHCO_3$) and 107.0 mL of 0.1 N NaOH, diluted to 1.000 L with water[b] or 500.0 mL of 0.025 M sodium tetraborate ($Na_2B_4O_7$) and 183.0 mL of 0.1 N NaOH, diluted to 1.000 L with water[b]

[a] Buffers were adjusted to exact pH by addition of acid or base, and pH measurements were made with Orion 701 A pH meter, Orion 91–03 pH electrode; SBC = standard buffer conditions

[b] Water used in preparation of buffers is purified by Milli-Q system (reverse osmosis, followed by two ion exchange filters, carbon filter, and 0.22-μm filter; resistance of water is usually > 10 Ωcm

[c] pH 10 buffer used instead of pH 11 buffer for ethyl acetate hydrolysis (see text)

Two important limitations on the procedures are: (1) compounds which ionize within this pH range may give rate constants which cannot be reliably extrapolated from one pH value to another by the method described and (2) measurements at elevated temperatures may lead to intervention by other chemical processes ordinarily too slow to be important at room temperature. Initial evidence of such processes usually can be detected by a change of products. Moreover, non-linear dependence of k on $e^{-1/T}$ for any single process also indicates intervention of some other process.

Reactivity and Structure-Activity Relationships

Reactivity

We can make some generalizations about the ranges of reactivity of specific classes and subclasses of compounds in pH 7 water at 25 °C. These generalizations are more usefully expressed in half-lives (t/2) than in rate constants and much of the information comes from Mabey and Mills' review [2].

Hydrolysis at saturated carbon (RX) generally increases with alkyl branching at the reaction center: methyl and ethyl bromide have t/2 of 25 days but isopropyl and t-butyl bromides have t/2 of 2 days and a few seconds, respectively. Similarly among leaving groups from RX the pK_a roughly predicts the best leaving groups as Cl, Br, and OSO_3Ph and the poorest as NHR and OR; the latter two are unreactive and the former have t/2 of days when R is n-alkyl and a few seconds if R is t-alkyl.

Carbonyl derivatives follow similar reactivity patterns as saturated carbon derivatives but with the important difference that in carbonyls, NHR and OR becomes moderately reactive leaving groups. Thus in the series RC(O)–X where X varies from Cl, F, OAc, OEt, to NH_2, t/2 varies from 0.8 s, 1.7 h, 22 min, 2 yrs, to 4,000 yrs, respectively.

Carbamates reactivity is complicated by two pathways for reaction in base (see below) but in the series PhNHC(O)–OR, t/2 varies with R in the same way as for carbonyls and alkyl compounds. When R is $p–NO_2Ph$, $m–ClPh$, $p–MeOPh$ or Me, t/2 is 26 s, 1 h, 1–3 days, and 4,000 yrs. If PhNH is replaced by NMe_2 in the above series then for $R = p–NO_2Ph$, t/2 is 3,000 yrs.

Substitution of X by OH at saturated carbon becomes much more facile if in addition to X there also is an OR or NR_2 substituent. Rapid hydrolysis of acetals and ketals in weak acid exemplifies this principle. In the series $ABC(OEt)_2$ as A and B change from A = B = H to A = H, B = Me to A = B = Me, t/2 changes from over 5,000 yrs to 300 days to 3 h. Orthoesters are much more reactive and exhibit less dramatic changes on additional substitution; in the series $AC(OEt)_3$, where A changes from H to Me to Ph, t/2 changes from 3.5 h to 6 min to 6 h [91]. α-chloroethers solvolyze with t/2 of a few minutes.

Epoxides are the exception to the rule that OR does not hydrolyze at saturated carbon, presumably because of relief of steric strain on ring opening. Epoxides with aliphatic substituents on the ring hydrolyze with t/2 of 10–20 days at pH 7; as with other groups, additional OR or NR_2 groups on the epoxide ring decrease t/2 to a few minutes.

Phosphorus esters exhibit complex dependence of hydrolysis on substitution, but the pK_a of the leaving group still plays an important role. In the series $(RO)_2P(O)X$, as X changes from Cl to OPh or OR, $t_{\frac{1}{2}}$ increases from 2 min to about 1 yr for both OPh and OR. If X is alkyl, one of the RO groups departs and t/2 is over 100 years.

Structure-Activity Relationships

Structure activity relationships (SARs) have been developed for many classes of hydrolyzable organic compounds. Indeed the original Hammett relationship [68] given in the form

$$\log k_x = \log k_o + \varrho\sigma \qquad (67)$$

was developed on the basis of a linear correlation between rate constants for hydrolysis of aromatic ethyl esters by base (k_B) and equilibrium ionization constants for corresponding aromatic acids (K_A). In equation (67) k_x and k_o are rate constants for substituted and unsubstituted aromatic structures, respectively, and σ is the substituent constant for X while ϱ is the measure of sensitivity of the reaction rate constant to changes in the substituent. The Hammett equation applies expressly to m- or p-substituted aromatic systems, where only electronic effects are important. For reviews of the subject see Chapman and Shorter [69, 70] and Jaffe [42].

Rates of hydrolysis for most organic chemicals are influenced by both steric and electronic effects of substituents in close proximity to or conjugated with the reaction center. Thus, hydrolysis of R–X to give R–OH is affected by steric and electronic effects in both the leaving group X and near the reaction center in R.

Hydrolysis of many classes of aliphatic carbonyl compounds fit the Taft equation [69–72]

$$\log k_x = \log k_o + \varrho \sigma^* + \delta E_s, \tag{68}$$

where in addition to the modified Hammett constants σ^*, there are additional terms for steric influence in substituents, E_s, and the sensitivity of the reaction to the steric effects, δ. The Taft equation is a valuable relationship for many classes of organic compounds for which the Hammett relationships fails owing largely to intervention of the steric effects at the reaction center.

SARs for hydrolysis also can be expressed in terms of a relation of the pK_a of the leaving group $-X$ to the rate constant for hydrolysis (Bronsted equation). Use of the pK_a values in an SAR implicitly assumes that loss of X is involved in the rate-controlling step for hydrolysis. Thus the ease of loss of $-X$ is related to the acidity of its conjugate acid and the SAR equation takes the form

$$\log k_x = \varrho pK_a + \log k_o. \tag{69}$$

The relation between the Hammett (67) and the Bronsted (69) pK_a equations follows from the correlation of equilibrium acidity constants in the Hammett form

$$\log K_x = \log K_o + \varrho \sigma, \tag{70}$$

$$pK_x = -\log K_x, \tag{71}$$

where K_x and K_o are acid dissociation constants for substituted and unsubstituted acids, respectively.

We must emphasize the importance of using SARs only in application to hydrolysis processes involving a single hydrolyzing agent and which follow a single reaction path. For example, hydrolysis of esters may involve hydrolysis by water, acids, or bases. The observed rate constant at some pH may be a combination of these individual processes, each of which has a different intermediate and transition state for which substituent effects will be manifested in different ways. Therefore, the composite rate constant may exhibit a complex or discontinuous relation between structure and reactivity. Failure of SARs generally arise from lack of correspondence between transition states among different compounds of a class owing to a change in reaction mechanism. Indeed the major application of SARs has been to elucidate transition state structures rather than predict rate constants for unmeasured new compounds [69, 70, 73].

In the following sections we will discuss briefly SARs for important classes of hydrolyzable compounds, the best correlation equations available for these reactions in water and ranges of reactivity. As has been noted elsewhere in this review a great many hydrolysis reactions have been studied quantitatively only in mixed organic water-solvents. Because solvent polarity sometimes can influence reaction rates significantly, though generally not predictably, these data are not very useful.

Carboxylic Acid Esters

Aliphatic esters can undergo acid, base, and neutral hydrolyses [2]. Acid catalyzed processes are relatively insensitive to electronic effects in the acyl group. The

Table 6. SARS for k_B for esters RC(O)OR' in water[a]

$$\log k_B = \varrho\sigma^* + \delta E_s + C$$

R	R'	ϱ	δ	C	n	r^2	Ref.
R	Et	2.49	1.05	−0.96	23	0.944	[2,77,81]
CH_3	R'	2.18	0.221	−0.96	16	0.943	[75,77,78,82,83]
R	R'	2.32(R) 2.10(R')	0.98	−0.74	37[b]	0.951	–
R	PhX	$1.21(\sigma)$[c]	0.618	+0.0792	24	0.986	–
XPh[d]	R	$2.18(\sigma)$[c]	0.73	−3.216	37	0.880	[75,83]
XPh	R	$2.58(\sigma)$[c,e]	1.92	−1.10	6	0.976	[75]

[a] σ^* values are from Perrin et al. [84]; E_s values are from MacPhee et al. [85] and $\sigma^* = 0$ for alkyl see [76]
[b] Cross correlation of two preceeding data sets with $\sigma^* = 0$ for alkyl groups
[c] Only X correlated since $\sigma^* = 0$ for alkyl groups
[d] Data set of Washkun et al. [83] measured in 50% AN−H_2O
[e] Only Me, Et, and i-Pr

neutral (water) process is important only when the acyl or alcohol groups are strongly electron-withdrawing. Base promoted hydrolysis (by OH$^-$) does exhibit significant variations with changes in structure of the acyl group.

For most esters the base – promoted reaction is the dominant process above pH 6 or 7 [2] although the neutral process can be important for many esters with electronegative substituents in the acyl group. Acid hydrolysis usually is dominant only below pH 4. The largest data base for hydrolysis kinetics constants in water is for values of k_B [2, 71, 74].

Recently Drossman and Mill [75] evaluated the published data and measured additional values of k_B for several kinds of esters (RC(O)OR') from which they developed several Taft and Hammett correlations for alkyl and aromatic esters. Table 6 summarizes these SARs.

Of special interest is the fact that ϱ values are similar for variations in R and R' but that ϱ values for XPh are much smaller in the alcohol than in the acid. Steric effects in acyl R or XPh become more significant as the size of the companion alcohol group (R') increases. However, the value of δ for alcohol R is small.

The last two equations correlate k_B for aromatic esters, in one case for the large data base in 50% acetonitrile [83] and in the other case for a far smaller group of the same esters in water [75]. Although R is not much changed in going to pure water, δ increases significantly. Reasons for this change are not clear and deserve validation through additional measurements in water.

The data base for purely aromatic esters in water is too small for meaningful correlation. Solvent effects for these processes are large: k_B for ethyl benzoate is 200 times smaller in 50% acetonitrile than in water [83,86].

The Taft correlations of Table 6 may be extended to more complex acyl groups by the use of additivity principles for σ^* and E_s. Mill et al. [86] have shown that σ^* for complex groups may be estimated reasonably well using additivity of simpler constituents for which σ^* values are known. For example in the complex structure XYZC, the value of σ^* for the group XYZC- is closely approximated by the sum

Table 7. SARs for carbamates[a]

ZYNC(O)OPhX

$$\log k_B = \varrho pK_a + C$$

Z	Y	ϱ	C
Ph	H	−1.15	13.6
Ph	Me	−0.26	− 0.26
Me	H	−0.91	9.3
Me	Me	−0.17	− 2.6

[a] [88]

of σ^* values for Ch_2X, Ch_2Y, and CH_2Z. Similarly, the value for E_s for the same group can be estimated roughly using the method of Austel et al. [87].

Aromatic Carbamates

Carbamates undergo acid, base, and neutral hydrolyses [2]. Wolfe et al. [88] extended and developed SARs for base-promoted hydrolysis of four types of carbamates in which k_B is correlated with the pK_a of the leaving phenol group

$$\log k_B = \varrho pK_a + C. \tag{72}$$

Table 7 summarizes SARs for those carbamates correlated by Eq. (72).

Two types of base-promoted mechanisms probably occur: (1) nucleophilic attack by OH^- on the carbonyl group occurs in all cases but is slow; (2) carbamates with N−H groups exhibit exceptional reactivity owing to reversible loss of a proton from nitrogen followed by cleavage to isocyanate and phenoxide and this process usually is more rapid than OH^- addition.

$$RNHC(O)OPhX \rightleftharpoons R\bar{N}C(O)OPhX, \tag{73}$$

$$R\bar{N}C(O)OPhX \longrightarrow RNCO + XPhO^-. \tag{74}$$

Large absolute values of Hammett ϱ [Eqn. (72)] for base-promoted hydrolyses of aromatic carbamates with CH_3NH groups indicates that the properties of the leaving phenoxide group are important in the rate-controlling step.

Phosphate Esters

Wolfe [89] evaluated SARs for hydrolysis of several phosphate and thiophasphates esters at 27 °C. These SARs correlated k_B and in some cases k_n with pK_a's of the alcohol or phenol leaving groups. The SARs take the same form as for carbamates (Table 7)

$$\log k_B = \varrho pK_a + C, \tag{75}$$

where ϱ is -0.24 ± 0.03 for $(RO)_2P(S)X$ or $(RO)_2P(O)X$ and X is a series of substituted phenols but also includes methoxy and ethoxy.

Hydrolysis of phosphate esters is complicated by the presence of up to three potential leaving groups. Thus the measured rate constant for hydrolysis of one of these leaving groups should take into account the favorable statistics for hydrolysis, where two or more identical leaving groups are present.

Rynard [90] has performed a series of molecular orbital calculations on phosphorus esters of the series RR'P(O)X, where R and R' are alkyl or alkoxy and X is F, Cl, or NO_2Ph and correlated those properties with $\log k_B$ [90]. The best global correlation of all data was achieved using a four-parameter equation including dipole moment, pK_a of X and E (HOMO).

Acetals and Ketals

Cordes [91] has reviewed the hydrolysis of acetals, ketals, and orthoesters, and provided a succinct listing of SARs for acid catalysed hydrolysis. Hammett and Taft relations correlate these data sets reasonably well and in one case, hydrolysis of aliphatic ketals and acetals [91], the correlation of σ^* with $\log k_A$ extended over a range of 10^7 with $\varrho^* = -3.6$ [91]. Relative hydrolysis rates for acetals and ketals increase by a factor of 10^4 on changing from CH_2 to CHMe and another 10^4 for the second Me in $CMe_2(OEt)_2$. The rate factor becomes 10^5 for substitution of Ph for H in this series [92].

The dominant pathway for hydrolysis of these compounds appears to involve formation of an intermediate carbenium ion which is rapidly intercepted by water with retention of configuration [91]. Unlike reactions of epoxides (below), where a water-promoted reaction is important, no significant reaction with water alone appears to be important for these cyclic ethers.

Epoxides

Hydrolysis of epoxides can be effected by acid neutral or base-promoted processes [2] but values of k_B usually are so small that water and acid processes dominate hydrolysis up to pH 10 or higher. No SARs have been published for the modest-sized data base that exists (2). Drossman and Mill [75] have measured pH-rate profiles for several substituted propylene oxides of the type $R_1HC{-}{-}{-}CHR_2$,

where R_1 is Me, $BrCH_2$, FCH_2, $PhCH_2$, and Cl_3C and R_2 is H or Me.

For most epoxides, the neutral process is close to $1 \times 10^{-6} s^{-1}$ with the exception of the Cl_3C epoxide for which k_N is 25 times larger.

The acid-catalysed process is much more sensitive to changes in R_1 or R_2, with a range of over 10^3 in k_A. Only fair correlations are found between $\log k_A$ and σ^* or σ^+ the special Hammett constant for electron-deficient reaction centers [75].

Alkyl Halides

Hydrolysis of alkyl halides was the center-piece of mechanistic organic chemistry during its formative period fifty years ago [1, 3, 93] and continues to exert a special fascination for chemists even today [1]. Alkyl and allylic halides hydrolyze by both

Table 8. General rules for alkyl halide hydrolysis

Type	k_h/s^{-1}			Reactivity factors
R–F	10^{-5}	to	10^{-10}	All except tertiary fluorides are quite stable
R–Cl	10^{-2}	to	10^{-8}	Chlorides and bromides are most
R–Br	10^{-1}	to	10^{-7}	reactive; bromides are 5–10 times
R–I	10^{-5}	to	10^{-8}	more reactive than chlorides
R–CH$_2$–X	10^{-3}	to	10^{-6}	S$_N$2, Highly sensitive to steric effects. Electronic effects are less important than in S$_N$1 mechanism
R$_2$CH–X	10^{-2}	to	10^{-5}	S$_N$2/S$_N$1. Sensitive to steric and
Ar–CR$_2$–X	10^{-1}	to	10^{-7}	electronic effects to varying degrees
R$_3$C–X	10^{-1}	to	10^{-4}	S$_N$1, Steric effects usually negligible (steric inhibition + steric acceleration). Special steric effects on cyclic halides. Highly sensitive to electronic effects
R$_2$C=CR–CH$_2$–X	10^{-4}	to	10^{-7}	Generally reactive
R–CH$_n$X$_{3-n}$	10^{-1}	to	10^{-12}	Polyhalomethanes (R=H) are very stable; aryl substitution can greatly enhance reactivity
$\diagdown$C=C$\diagup$ $\quad$X	$< 10^{-10}$			Very stable
Ar–X	$< 10^{-10}$			Extremely stable unless electron-donating substituents in ring
ArCH$_n$X$_{3-n}$	10^{-5}	to	10^{-2}	Very reactive
R or ArC(O)X	$> 10^{-2}$			Very reactive

neutral and base-promoted reactions [2] to give mostly alcohols and in some cases small amounts of olefins, which arise from base-promoted elimination in polyhaloalkanes or from solvolysis of t-alkyl halides [1, 3]. The mechanistic features of these reactions were addressed in the second section.

Significant structural effects on rates of hydrolysis of RX reside both in R and in X but usually are limited to close proximity to the reaction center. Quantitative SARs for alkyl halides generally are lacking but for a given X rate constants increase in order of increasing alkyl or alkoxy substitution, increasing by ten-fold for one additional methyl and by 100,000 for an alkoxy substituent. Halogen substitution dramatically inhibits hydrolysis [2]. Rate constants also increase in the series F, Cl, Br, I with the largest change occurring on changing from F to Cl.

Table 8 summarizes the ranges of rate constants for most important classes of alkyl halides together with some useful generalizations about reactivity.

Most alkyl halides hydrolyze with rate constants which are independent of pH below pH 10 and with t/2 of 10 to 40 days for n- or iso-alkyl groups; t-BuCl has a t/2 of just 25 s, however. Benzyl halides have t/2 of ranging from a few minutes to several hours. Allylic halides are less reactive with t/2 of several days while polyhalomethanes and ethanes react with t/2 generally in excess of 100 years.

Hydrolysis of vinyl, aromatic, neopentyl, and bridgehead halides usually are so slow that reliable rate constants have not been measured in water near 25 °C.

Other SARs

For number of other classes of organic chemicals, SARs are available which describe limiting ranges of rate constants as a function of structure. These include alkyl halides, amides, anhydrides, and isocyanates [40, 74]. In the case of the latter two classes of compounds hydrolysis rates generally are so rapid that for practical fate or exposure assessment the compounds will disappear in just a few hours on dissolution in natural waters. Other classes such as alkyl or aromatic amides are generally so unreactive under ordinary conditions that they may be considered persistent. Epoxides fall in between in having relatively little sensitivity of structure to the rates of the water-promoted reaction (k_n) and generally exhibit half lives of about 10–20 days in natural waters.

Qualitative SARs such as these are immensely valuable in focusing attention on important compounds or classes of compounds for exposure assessment and eliminating the need to further consider extremely reactive or unreactive classes.

References

1. For a general but detailed review of hydrolysis and related reactions in a mechanistic framework see Lowry, T.H., Richardson, K.S.: Mechanism and Theory in Organic Chemistry. Harper and Row, New York 1981
2. Mabey, W.R., Mill, T.: J. Phys. Chem. Ref. Data 7, 383 (1978)
3. Streitweiser, A.: Solvolytic Displacement Reactions, McGraw-Hill, New York 1962
4. Thornton, E.R.: Solvolytic Mechanisms. Ronald Press, New York 1964
5. Bamford, C.H., Tipper, C.F.H. (eds.): Comprehensive Chemical Kinetics, Vol. 10, Elsevier, Amsterdam 1972
6. Harris, J.R.: Prog. Phys. Org. Chem. 11, 89 (1974)
7. Bentley, T.W., Schleyer, V.R.: Adv. Phys. Org. Chem. 14, 1 (1977)
8. Gleave, J.L., Hughes, E.D., Ingold, C.K.: J. Chem. Soc. 236 (1935)
9. Patai, S. (ed.): The Chemistry of the Carbonyl Group, Vol. 1, Wiley, New York 1966
10. Fife, T.H.: Adv. Phys. Org. Chem. 11, 1 (1975)
11. Jencks, W.P.: Prog. Phys. Org. Chem. 2, 63 (1964)
12. Bender, M.L.: Chem. Rev. 60, 53 (1960)
13. Satchell, D.P.: J. Chem. Soc. 558 (1963)
14. Ingold, C.K.: J. Chem. Soc. 1032 (1930)
15. DeTar, D., Tenpas, C.J.: J. Amer. Chem. Soc. 98, 7903 (1977)
16. Taft, R., Talbot, R.J.E.: In Comprehensive Chemical Kinetics, Bamford, C.H. and Tipper, C.F.H. (eds.), Vol. 10. Elsevier, Amsterdam 1972
17. Johnson, S.L.: Adv. Phys. Org. Chem. 5, 237 (1967)
18. Whalen, D.L.: J. Amer. Chem. Soc. 95, 3432 (1973)
19. Swain, C.G., Scott, C.B.: J. Amer. Chem. Soc. 75, 141 (1953)
20. Bartlett, P.D., Swain, C.G.: J. Amer. Chem. Soc. 71, 1406 (1949)
21. Fife, T.H., DeMark, B.R.: J. Amer. Chem. Soc. 101, 7379 (1979)
22. Garrett, E.R., Goto, S.: Chem. Pharm. Bull. 21, 1811 (1973)
23. Tsuji, A., Yamana, T., Mizukani, Y.: Chem. Pharm. Bull. 20, 2528 (1972)

24. Bender, M.L., Chow, Y.-L., Chloupek, F.: J. Amer. Chem. Soc. *80*, 5380 (1958)
25. Jencks, W.P.: Accts. Chem. Res. *9*, 425 (1976)
26. Hegarty, A.F., Hegarty, C.N., Scott, F.L.: J. Chem. Soc., Perkin II, 1258 (1974)
27. Kirby, A.J., Mujahid, T.G., Camilleri, P.: J. Chem. Soc., Perkin II, 1610 (1979)
28. Fife, T.H., Duddy, N.W.: J. Amer. Chem. Soc. *105*, 74 (1983)
29. Fedor, L.R., Bruice, T.C.: J. Amer. Chem. Soc. *87*, 4138 (1965)
30. Bruland, K.: In Chemical; Oceanography, Vol. 18, Riley, J.P., Skirrow, G. (eds.), Academic Press, London 1983
31. Quality of Surface Water in the United States, U.S. Geological Survey Water Supply Papers 2091–2100, 1973
32. Christman, R.F., Gjessing, E.T. (eds.): Aquatic and Terrestrial Humic Materials. Ann Arbor Science, Ann Arbor MI 1973
33. Pytkowicz, R.M., Kester, D.R.: Oceanogr. Mar. Biol. Ann. Rev. *9*, 11 (1971)
34. Blessing, J., Mill, T.: Unpublished summary of data in [31], 1976
35. Horne, A.: The Chemistry of Our Environment. Wiley, New York 1978
36. Florence, T.M., Bately, G.E.: Chemical Speciation in Natural Waters, CRC Critical Reviews in Analytical Chemistry, CRC Press. Boca Raton FL 1980
37. Stumm, W., Morgan, J.J.: Aquatic Chemistry, 2nd ed., Wiley, New York 1981
38. Perdue, E.M., Wolfe, N.L.: Environ. Sci. Technol. *17*, 635 (1983)
39. Frost, A.A., Pearson, R.G.: Kinetics and Mechanism, 2nd ed., Wiley, New York 1961
40. Mabey, W.R., Drossman, H., Liu, A.M., Mill, T.: Toxic Substances Process Data Generation, EPA Final Report, EPA Contract 68-03-2921, Task 18, 1984
41. Robertson, R.E., Scott, J.M.W.: J. Chem. Soc. 1596 (1961)
42. Jaffe, H.H.: Chem. Rev. *53*, 191 (1953)
43. Martell, A.E.: In Organic Compounds in Aquatic Environments, Faust, S.J., Hunter, J.V. (eds.). Dekker, New York 1971
44. Hoffman, M.R.: Environ. Sci. Technol. *14*, 1061 (1980)
45. Florence, T.M.: Water Res. *11*, 681 (1977)
46. Buckingham, D.A., Clark, C.R.: Austral. J. Chem. *35*, 531 (1982)
47. Buckingham, D.A., Englehardt, L.M.: J. Amer. Chem. Soc. *97*, 5915 (1975)
48. Ketalaar, J.A., Gersmann, H.R., Beck, M.M.: Nature *122*, 392 (1956)
49. Steffans, J.J., Sampson, E.J., Siewers, I.E., Benkovic, S.J.: J. Amer. Chem. Soc. *95*, 936 (1973)
50. Blanchet, P., St. George, A.: Pestic. Sci. *13*, 85 (1982)
51. Meikel, W.R., Youngson, C.R.: Arch. Environ. Contam. Toxicol. *7*, 13 (1978)
52. Mortland, M.M., Raman, K.V.: J. Agric. Food Chem. *15*, 163 (1967)
53. Mabey, W.R., Baraze, A., Mill, T.: In Environmental Pathways of Selected Chemicals, Part I, EPA Final Report, EPA 600/7-78-074, 1978
54. Karickhoff, S.W.: J. Hydraulic Eng. *110*, 707 (1984)
55. Chiou, C.T., Peters, L.J., Freed, V.H.: Science *206*, 831 (1979)
56. Karickhoff, S.W., Brown, D.S., Scott, T.A.: Water Res. *13*, 241 (1979)
57. Yalkowski, S.H., Valvani, S.C.: J. Pharm. Sci. *69*, 912 (1980)
58. Macalady, D.L., Wolfe, N.L.: Amer. Chem. Soc. Sympos. Ser. *259*, 221 (1984)
59. Sethunan, N., MacRae, I.C.: J. Agr. Food Chem. *17*, 221 (1969)
60. Mingelgrin, U., Saltzman, S., Yaron, B.: Soil Sci. Soc. Amer. J. *41*, 519 (1977)
61. El-Amamy, M.M., Mill, T.: Clays Clay Miner. *32*, 67 (1984)
62. Choudhry, C.G.: Humic Acids
63. Wereshaw, R., Goldberg, M.: Adv. Pest. Chem. *111*, 149 (1972)
64. Carter, C.W., Suffet, I.H.: Environ. Sci. Technol. *16*, 733 (1982)
65. Landrum, P.F., Nilhari, S.R., Eadle, B.J., Gardner, W.S.: Environ. Sci. Technol. *18*, 187 (1984)
66. Mill, T., Mabey, W.R., Smith, J.H., Bomberger, D., Chou, T.-W.: Laboratory Protocols for Environmental Processes, EPA Final Report, EPA 600/3-82-022, 1982
67. Harned, W., Owen, G.: Physical Chemistry. Reinhold, New York 1958
68. Hammett, L.P.: J. Amer. Chem. Soc. *59*, 96 (1937)
69. Chapman, N.B., Shorter, J. (eds.): Advances in Linear Free Energy Relationships. Plenum, London 1972
70. Chapman, N.B., Shorter, J.: Correlation Analysis in Chemistry. Plenum, London 1978
71. Hine, J.: Structural Effects on Equilibria in Chemistry. Wiley, New York 1975

72. Taft, R.W.: J. Amer. Chem. Soc. *75*, 4231, 4534, 4538 (1953)
73. Exner, O.: In [70]
74. Kirby, A.J.: In Comprehensive Chemical Kinetics, Vol. 10. Bamford, C.H., Tipper, C.F.H. (eds.). Elsevier, Amsterdam 1972
75. Drossman, H., Mill, T.: SRI, unpublished results
76. DeTar, D.: J. Org. Chem. *45*, 662 (1980)
77. Salmi, E.J., Leimu, R.: Suomen Kemist. *20B*, 43 (1947)
78. Euranto, E.K., Moisio, A.: Suomen Kemist. *37B*, 92 (1964)
79. Bell, R.P., Collier, B.A.W.: Trans. Farad. Soc. *61*, 1445 (1965)
80. Hay, R.W., Porter, L.J.: J. Chem. Soc. 1261 (1961)
81. DePuy, C.H., Mahoney, L.R.: J. Amer. Chem. Soc. *86*, 2653 (1964)
82. Bogatov, S., Cherkasova, E.: Organic React. (Tartu) *14*, 165 (1965)
83. Washkun, R.J., Patel, V.K., Robinson, J.R.: J. Pharm. Sci. *60*, 736 (1971)
84. Perrin, D.D., Dempsey, B., Serjeant, E.P.: pKa Prediction for Organic Acids and Bases. Chapman and Hall, New York 1981
85. MacPhee, J.A., Panaye, A., Dubois, J.: Tetrahedron *34*, 3553 (1978)
86. Mill, T. et al.: Data Generation in Process and Property constants used in Fate Assessment, EPA Final Report, EPA Contract No. 68-03-2981, 1984
87. Austel, V., Kutter, E., Kalbfleisch, W.: Arnzeim. Forsch. (Drug Res.) *3*, 585 (1979)
88. Wolfe, N.L., Zepp, R.G., Paris, D.F.: Water Res. *12*, 561 (1978)
89. Wolfe, N.L.: Chemosphere *9*, 571 (1980)
90. Johnson, H., Kenley, R.A., Rynard, C.: Investigation and Design of Simulant Molecules for Organophosphorus Esters, NASA Final Report, Contract NAS2-11249, 1983
91. Cordes, E.H.: Prog. Phys. Org. Chem. *4*, 1 (1967)
92. Kreevoy, M., Taft, R.W.: J. Amer. Chem. Soc. *77*, 5590 (1955)

Exchange of Pollutants and Other Substances Between the Atmosphere and the Oceans

Michael Waldichuk

West Vancouver Laboratory, Department of Fisheries and Oceans
West Vancouver, B.C. V7V 1N6, Canada

Summary

The atmospheric input of some substances to the oceans can exceed the riverine input, e.g., by a factor of 10 for Pb. Emissions to the atmosphere of Pb, Zn, Mn, and Cu have been greater from anthropogenic than from natural sources, i.e., ocean, earth's crust, volcanoes, and vegetation. Transfer of substances from the atmosphere to the sea occurs mainly by rain washout, but also as dry fallout. Sea-to-atmosphere transfer is largely by bubble bursting. The thin (< 100 μm) sea-surface microlayer is the site of enrichment of substances coming from both the atmosphere and the seawater column. Substances with an anthropogenic component increase in atmospheric concentration from oceanic areas to urbanized regions. For example, estimated mean fluxes of Cu and Zn at Enewetak were 2 and 13 ng cm^{-2} yr^{-1} and in the North Sea 1,300 and 8,950 ng cm^{-2} yr^{-1}, respectively.

The net flux of anthropogenic CO_2, estimated as 2×10^{15} g C yr^{-1}, is from the atmosphere to the sea. An annual, global net flux from the ocean to the atmosphere of certain biogenic gases is estimated, as follows: CO, 10^{14} g; CH_4, 10^{12}–10^{13} g; CH_3I, 3×10^{11} g; $(CH_3)_2S$, 30–50 $\times 10^{12}$ g; and N_2O, 6×10^{12} g. Methane is now considered to be an important "greenhouse" gas. The annual production of wind-blown dust has been estimated at 5×10^{14} g, and the seasonal pattern in the dust flux at Pacific oceanic islands is related to the incidence of dust storms in Asia. The annual sea-to-air flux of marine sea-salt aerosols has been estimated at 3×10^{15} g, or 1×10^{15} g of Na. Seasonal global distributions of aerosols calculated from mean surface winds, show that concentrations of 10–15 μg m^{-3} are typical year-round for the low-wind zone from 30° S to 30° N, but may exceed 40 μg · m^{-3} in windy areas at high latitudes during winter.

Among problems in air-sea flux evaluations are measurements of dry fallout and proportion of the net deposition of trace elements associated with recycled sea spray. The SEAREX (Sea-Air Exchange) Program is making significant advances in better understanding sea-air exchange processes in order to obtain more reliable material flux data.

Introduction

During the last two decades, two major environmental issues have emerged that concern atmospheric emissions of anthropogenic substances and their transfer from the atmosphere to water. One involves the freshwater environment and its acidification by sulphur dioxide and oxides of nitrogen emitted mainly by smelters and fossil-fuel burning power plants. The other involves the increasing atmospheric temperature expected from the greenhouse effect being created by excessive carbon dioxide emitted into the atmosphere by fossil-fuel combustion. In the first case, the ecological problems are largely confined to the terrestrial environment, particularly lakes and rivers that are poorly buffered with weak-acid salts. The effect on the oceans of these acid-rain precursors (e.g., SO_2, NO_x) would be rather minimal, as we understand it today, because of the high buffering capacity of seawater. In the second case, the effect of increasing atmospheric temperature on the oceans would be mostly indirect, in that sea level would rise from melting of Arctic and Antarctic ice. There would be, of course, more subtle changes arising from increasing temperature of the seawater itself. But the importance of the oceans with respect to the increasing atmospheric CO_2 is in the role they play in absorbing the excess CO_2 and thus buffering its effect in the atmosphere. It is not intended to delve too deeply into these major issues in this chapter, because they have been reviewed quite adequately elsewhere [1–4], but only to provide some highlights relevant to the general topic of air-sea exchange of pollutants.

It is now recognized that a substantial proportion of materials entering the sea from land is transferred through the atmosphere [3,5–7]. This is particularly true for substances emitted directly into the atmosphere, such as SO_2 and NO_x from smelters and coal-fired power generating plants. But it also applies to substances not intended for atmospheric dispersal, as for example, pesticides used in agriculture and forestry [8, 9], and metals from various sources. Even materials that are normally confined and do not usually have atmospheric contact, e.g., polychlorinated biphenyls (PCBs) [8], may undergo leaks and accidental spills, evaporate and may be transferred to the marine environment by way of the atmosphere.

We normally think of a one-way flux, that from the atmosphere to the oceans, when we consider air-sea exchange of substances. There are various materials, mostly of natural origin, that are involved in an ocean-to-atmosphere flux. These substances introduce a certain amount of uncertainty in mass-balance calculations inasmuch as they may have both anthropogenic and natural sources. Aerosols above the sea surface and above the adjoining land are virtually totally derived from the sea.

As noted above, the transfer of some substances between the atmosphere and the sea, e.g., SO_2, may have little environmental consequence in the oceans because seawater is well buffered chemically. There can be a negative impact, however, from the atmospheric input of such substances as radionuclides, DDT, lead and mercury on the marine biota, particularly that in the sea-surface microlayer. Likewise, a negative impact can arise from sea-to-atmosphere transfer of pathogenic organisms to human populations from sewage-polluted seawater, for example.

It is intended to review in this paper the sources of substances that enter into air-sea exchange, the processes that are responsible for transfer of such substances, some indication of the global fluxes of selected substances, and a perspective of the environmental impact of exchanged substances. There will be some overlap with other chapters in these volumes in the material presented. This is unavoidable if the subject matter is to be reasonably well covered for a cohesive story. But an effort has been made to minimize this by giving only information essential for this chapter in parts where such overlap may occur. Radionuclides are not covered here, even though they were largely responsible for the initial public awareness of tropospheric transport of man-made substances. The subject of air-sea exchange of radioactive materials is highly specialized and considered beyond the scope of this chapter.

Sources of Substances for Air-Sea Exchange

It is difficult to separate material entering the sea as to whether it comes from land via rivers or the atmosphere or whether it is produced *in situ*. We do know that much of the organic material transported toward the sea in rivers is trapped in the estuaries [10]. The same applies, at least in part, to inorganic materials transported by rivers. It is becoming increasingly apparent that the amount of metals transported to the sea from land by the atmosphere is of the same order of magnitude as that transported by rivers (Table 1). Organic substances produced biogenically in the ocean can sometimes be differentiated from the anthropogenic contribution by isotopic carbon ratios.

Table 1. Ratio of atmospheric flux to riverine flux (dissolved) of some trace elements in coastal and semi-remote oceanic areas (From [6])

Element	South Atlantic Bight [11]	New York Bight [12]	North Sea [13]	Western Mediterranean Sea [14]
As	2.1	1.0	1.7	–
Cd	2.7	3.1	1.1	–
Cu	1.9	–	1.9	–
Fe	5.8	6.4	1.7	–
Mn	0.6	–	0.8	–
Hg	22	–	2.1	0.8
Ni	1.7	–	1.3	–
Pb	9.5	20	6.8	6.2
Zn	2.3	3.1	1.9	0.8

No attempt is made here to provide accurate quantitative information on global sources of substances available for air-sea exchange. While some effort has been made regionally, such as in the North Sea under ICES (International Council for the Exploration of the Sea) [15] auspices, to obtain contaminant input data, there are no reliable global input data.

Terrestrial

Accelerated efforts have been made in recent years to obtain quantitative data of inputs of contaminants to the atmosphere from such sources as smelters and coal-fired power plants, not so much from the marine point of view as from the freshwater acidification aspect. The amount of CO_2 carbon injected into the atmosphere by fossil fuel burning globally has been reasonably well established at about 5×10^{15} g per year currently [16]. The biotic, e.g., deforestation, contribution of CO_2 has not been nearly as well established [16, 17]. In Europe, various regional bodies such as ICES [15] and the Helsinki Commission have sought atmospheric input data for studies of contamination of the North and Baltic seas. A valuable contribution in this regard was the paper by Pacyna and coworkers [18] that provided data on the atmospheric input of metals from European states. It should be borne in mind, however, that the atmospheric burden of such substances as metals consists of both anthropogenic and natural inputs. Sometimes a neglected natural input to the atmosphere of not only organic but also inorganic constituents is vegetation. Evidence has been presented that particles containing high concentrations of zinc can be released into the atmosphere by growing plants [19], and a marine biogenic agent was proposed as the source of 20,000-fold enrichment of copper during aerosol production near the island of Tasmania [20]. Volcanoes have been suggested as a significant source of trace metals in the atmosphere [21–23].

Marine

Sea-salt particles make up a large proportion of the particulate material in the global marine atmosphere on a mass and volume basis. The annual global flux of sea-salt particles from the oceans to the atmosphere has been estimated at 10^{15} g by Eriksson [24]. This estimate was increased by a factor of 10 by Blanchard [25]. Concentrations of atmospheric sea-salt particles can be regarded as a measure of the amount of other marine substances injected into the atmosphere by sea-surface processes, inasmuch as the sea-salt aerosol serves as a vehicle for various trace species. The amount of anthropogenic substances injected from seawater into the atmosphere will probably bear some relationship to the amount of sea-salt introduced. Based on the flux of sea-salt particles to the atmosphere estimated by Eriksson [24], the Workshop on the Tropospheric Transport of Pollutants to the Ocean, held in Miami, Florida, U.S.A., December 1975 [3] evaluated the global flux of metals to the atmosphere. This is shown in Table 2 along with the calculated global flux of metals to the atmosphere from crustal weathering, based on a total crustal material flux of 2.5×10^{14} g $\cdot$ yr^{-1} estimated by Goldberg [26], the crustal abundances of metals given by Taylor [27], and the global annual emission of metals from fossil fuel combustion estimated by Bertine and Goldberg [30].

It is of some interest to see how the above estimates endured with time as more data were acquired. The Bubble Interfacial Microlayer Sampler (BIMS) was used

Table 2. Global flux of metals to the atmosphere based on total crustal material flux of 2.5×10^{14} g yr^{-1}[a] and total sea-salt flux of 1×10^{15} g yr^{-1}[b] (From [3])

Element	Crustal material[c] 10^9 g yr^{-1}	Bulk sea-salt[d] 10^9 g yr^{-1}	Fossil-fuel combustion products[e] 10^9 g yr^{-1}
Al	20,000	0.15	1,400
Fe	14,000	0.5	1,400
Na	6,000	3×10^5	300
Mn	200	0.005	7
Sc	6	0.000015	0.7
Cu	14	0.04	2
V	30	0.05	12
Se	0.013	0.003	0.5
Pb	3	0.0008	150[f]
Cd	0.05	0.0008	–
As	0.5	0.05	0.7
Zn	18	0.08	0.5
Sb	0.05	0.007	–
Hg	0.02	0.0005	1.6

[a] Goldberg [26]
[b] Eriksson [24]
[c] Using crustal abundances of Taylor [27]
[d] Using seawater concentrations of Riley [28] and Chester and Stoner [29]
[e] From estimates of Bertine and Goldberg [30]
[f] Estimate of Patterson et al. [31]

Table 3. Concentration of metals in seawater (From [32])

Element	Ocean	Concentration $\mu g\, kg^{-1}$	Reference
Al	Atlantic	0.5	[33]
Co	Indian	4×10^{-2}	[34]
Cu	Atlantic	1×10^{-1}	[35, 36]
Cu	Pacific	6×10^{-2}	[37]
Fe	Pacific	6×10^{-2}	[38]
K	Global average	3.8×10^{5}	[39, 40]
Mg	Global average	1.4×10^{6}	[39, 40]
Mn	Atlantic	1×10^{-1}	[41, 36]
Mn	Pacific	6×10^{-2}	[42]
Na	Global average	1.1×10^{7}	[39, 40]
Pb	Atlantic	3.5×10^{-2}	[33]
Pb	Pacific	1×10^{-2}	[43]
Sc	Atlantic	6×10^{-4}	[39]
V	Atlantic	1.2	[33]
V	Pacific	1.5	[44]
Zn	Atlantic	4×10^{-3}	[36]
Zn	Pacific	5×10^{-3}	[45]

in the North Atlantic to determine the enrichment factor, EF_{sea} (the ratio of the concentration of any element to sodium in the aerosol particle divided by the corresponding ratio in bulk seawater), for a series of trace metals on bubble-derived sea-salt particles [32]. It was assumed that the mean BIMS EF_{sea} values measured in the North Atlantic off Bermuda were applicable globally. Literature values were used for trace metal concentrations in surface seawater (Table 3), and for the total annual transfer of sea-salt to the atmosphere, ranging from 1×10^{15} g to 1×10^{16} g [24, 25, 46, 47]. A geometric mean of the reported annual flux of sea-salt to the atmosphere, 3×10^{15} g, was used, and this is equivalent to a transfer of 1×10^{15} g of sodium from the ocean to the atmosphere per year. The calculated global fluxes of metals from the oceans to the atmosphere, using BIMS data, are shown in Table 4, with the emission from other natural and anthropogenic sources given for comparison.

The contributions of metals from crustal weathering given in Table 4 were calculated by multiplying the mean crustal concentrations of the metals [56] by an estimate of annual global production of wind-blown dust, 5×10^{14} g [53]. Estimates of the emissions of metals from volcanoes were made by multiplying the trace metal-to-aluminum ratio in particulate matter in volcanic plumes and volcanic lava [57, 21, 58] by the total quantity of aluminum emitted annually by volcanoes, 7×10^{11} g. This calculation assumes that the annual production of particulate material by volcanoes is 1×10^{13} g [53] and that the Al content of volcanic material is the same as that of crustal material. If measurements made on particles from the hot vents of Mt. Etna [20], which gave much greater ratios of all metals to Al, are typical of input by volcanoes globally, then the values for some metals given in Table 4 should be higher by 1 or 2 orders of magnitude. The input of trace metals to the atmosphere by vegetation was calculated from the

Table 4. Source strengths for atmospheric trace metals (10^9 g · yr^{-1}) (From [32])

Element	Natural						Anthropogenic		Reference[b]
	Ocean (BIMS)[a]		Crust	Volcanoes	Vegetation		Fossil fuel	Other	
	Mean	1-m							
Al	200		20,000	700	40		2,000	2,000	[51]
Co	0.2		7	0.1	0.04		0.9	2	[52]
Cu	5		10	6	0.9		2	50	[53]
K	30,000		6,000	200	100		200	200	[51]
Fe	50	200	10,000	300	20		2,000	4,000	[51]
Mg	100,000		3,000	80	20		300	200	[51]
Mn	7		200	9	5		8	400	[54]
Pb	8	20	3	0.4	0.2		4	400	[53]
Sc	0.0005	0.005	7	0.2	0.002		0.8	0.4	c
V	10		30	0.7	0.2		20	2	[55]
Zn	8	100	80	10	10		80	200	[53]

[a] BIMS = Bubble Interfacial Microlayer Sampler [48–50]
[b] References are for non-fossil fuel anthropogenic sources
[c] Assumed crustal ratio to Fe emissions for ore values

concentrations of metals in dried plant tissue [59, 60] and total annual emission of particles from vegetation, 7.5×10^{13} g [53]. Not only is there much uncertainty in the latter figure, but the calculations assume that metals are not preferentially enriched on particles emitted by vegetation, and that there is no variation in metal composition of plant tissues.

The fossil-fuel component of the anthropogenic contribution of metals to the atmosphere was evaluated by multiplying the concentrations of metals in fossil-fuels [30] by the annual combustion of coal and petroleum, 3.1×10^{15} g and 2.8×10^{15} g, respectively, [53]. The other anthropogenic contribution of metals to the atmosphere comes mainly from mining and smelting of ores, manufacturing, and gasoline additives. Literature values were used directly for these emissions, or if unavailable, they were calculated from the production rate of each industry, emission factors [61] and the metal concentrations in the emissions.

When one compares data given in Table 4 with those evaluated a decade earlier, given in Table 2, it is impressive to note how the global sea-salt flux values particularly have changed. They have all increased by factors of: Al, 1333; Fe, 100; Mn, 1400; Sc, 33; Cu, 125; V, 200; Pb, 10,000; and Zn, 100. Changes in the values for metal contributions from weathering of crustal material and from fossil-fuel combustion have not been as drastic over the 10-year interval and the direction of change has not always been the same. The reason for this is probably that the really new data are for metals from the ocean source, derived with the BIMS, whereas the data for metals from other sources stem from evaluations using the same or similar estimates in both cases. It was recognized at the time Table 2 was compiled [3] that if significant fractionation of the trace metals occurs during bubble bursting, the fluxes calculated from bulk sea-salt concentrations would be too low. Moreover, it was considered at that time that estimates of metal ejection into the atmosphere from anthropogenic sources were probably more accurate than from crustal weathering, the ocean, or any other natural source.

Weisel et al. [32] demonstrated by their BIMS measurements of oceanic contribution of metals to the global atmosphere, and by their estimates of inputs from other sources that the ocean is the prime source of atmospheric K and Mg. The ocean may also account for 5–20% of the global input of Cu, V, and possibly Zn to the atmosphere. It is an insignificant contributor, however, of Al, Co, Fe, Mn, Pb, and Se. During certain times of year, in the absence of aeolian dust, in remote parts of the North Pacific, the ocean may be the principal source of all the trace metals present in aerosol particles greater than ~ 3 μm in diameter. Mean EF_{sea} values obtained for samples of sea-salt particles produced by the BIMS in the North Atlantic [32] were: $K = 1.0$; $Mg = 1.0$; $Sc \cong 10$; $Co \cong 60$; $Cu \cong 800$; $Pb \cong 4,000$; $Mn \cong 1,000$; $Al \cong 5,000$; $Fe \cong 10,000$; and $Zn \cong 20,000$. For Fe, Pb, Sc and Zn, EF_{sea} values increased as the bubble generation depth was increased, which suggests that these metals are scavenged by rising bubbles. But no such increase in EF_{sea} values was observed for K and Mg as the bubble generation depth was increased.

Because of the many assumptions involved, some of which may be only partially valid, in the evaluations of Weisel et al. [32] for the flux of trace metals from the oceans to the atmosphere, the flux values given in Table 4 can only be regarded as order-of-magnitude estimates.

The natural contribution of substances to the atmosphere from the oceans varies geographically according to a number of factors, such as biological activity and upwelling, in addition to winds and wave action. The worldwide variation of input of both natural and anthropogenic substances to the atmosphere from the sea is not known, but more and more information is being obtained in both regional programs (e.g., ICES) and oceanic studies under the SEAREX (SEA-AIR EXCHANGE) Program [62, 63, 7].

Processes and Evaluations of Transfer of Substances Between the Atmosphere and the Sea

The processes involved in the transfer of substances between the atmosphere and the sea are diagrammatically illustrated in Fig. 1.

Atmosphere-to-Sea Transfer

In general, substances originating in terrestrial areas are transported seaward by tropospheric wind systems, extending from the earth's surface to an elevation of 10–15 km, where the tropopause creates a discontinuity between the troposphere and stratosphere. In situations where substances are carried aloft into the stratosphere (10–15 km to 50 km), they become airborne on stratospheric winds and can be transported virtually around the globe. During wind-borne transport,

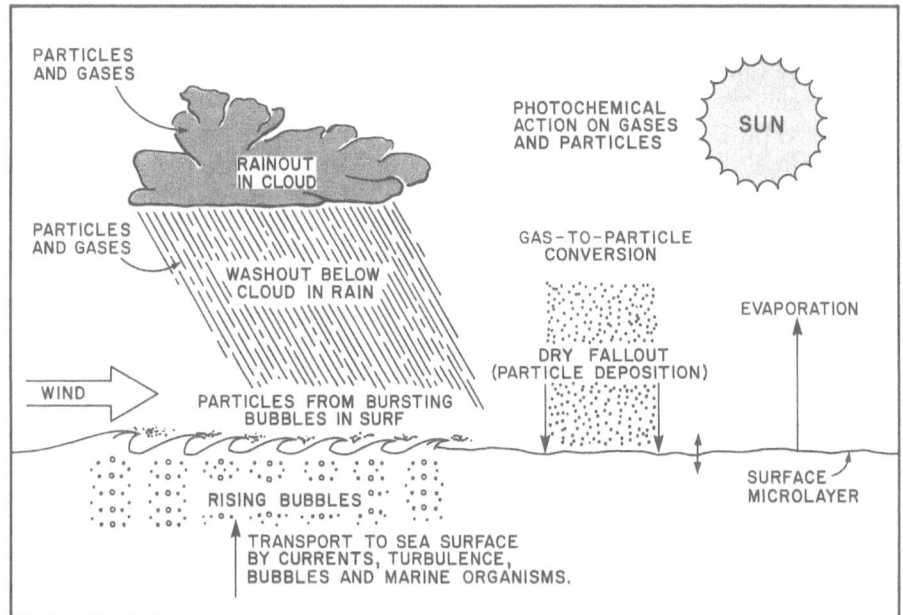

Fig. 1. Schematic diagram illustrating the processes involved in the air-sea exchange of substances

substances are subjected to a number of physicochemical processes, which may affect their form, and to certain physical effects that lead to their deposition in the sea.

Particulate Matter

Particulate matter varies in size of individual particles and in density. The larger the particles and the greater their density the more rapidly they will settle because of gravitation. The finer and lighter the particles the more likely they will be kept aloft by air currents. Brownian motion can also be a relatively important mechanism for their atmospheric movement. In any event, a certain proportion of the particles will settle to the sea surface by gravitation in the form of dry fallout.

Dry deposition of aerosol particles to the sea surface includes all deposition processes except precipitation. Estimation of dry deposition can be made by use of the deposition velocity, V_d, given by

$$V_d = F/M, \qquad (1)$$

where F is the flux of particles to the surface in $g \cdot cm^{-2} \cdot sec^{-1}$, and M is the atmospheric concentration in $g \cdot cm^{-3}$. It is not technically feasible at present to make direct measurements of dry deposition of trace substances on aerosol particles to the ocean surface. A reasonably accurate estimate of dry deposition can be made by applying available models relating dry deposition to the water surface as a function of particle size [64–66]. Use of such models, however, requires an accurate measurement of the mass-size distribution of the trace metals. Moreover, deposition velocities depend strongly on particle size, relative humidity and wind speed. Dry deposition calculations of sea-salt aerosol particles have been rather successfully carried out, provided that air samples were taken that represented the true sea-salt particle mass and size distributions [67].

Inasmuch as direct measurements of dry deposition to the ocean of such elements as Pb, Cd, and As are not possible, and model calculations suffer various shortcomings, most data for dry deposition at present are derived from use of surrogate surfaces, such as plates, filters or open vessels, for deposition measurements [e.g., 63, 68]. There has been some criticism of the use of such surfaces, because they are not considered to accurately mimic the characteristics of natural water surfaces [69].

Real biases are introduced by use of surrogate surfaces, such as a rimless plastic plate, but model calculations indicate that dry deposition to the sea surface and to such a plastic plate should agree within a factor of 2 or 3 over a broad range of wind speeds and relative humidities. This agreement has been shown to hold for sea-salt aerosol and mineral aerosol particles over the North Pacific [63]. The correct magnitude of total dry deposition of some trace metals may be inferred from use of such sampling devices, provided the agreement noted above holds for sub-micrometer aerosol particles. But measurements with such sampling devices should be made concurrently with the determination of the mass-size distribution of the elements of interest. For representative annual dry deposition data at a given location, sampling should be conducted at various times of the year. Unfortunately, few such data are available outside the SEAREX studies at the Pacific island stations [62, 63].

Airborne particles may also be removed from the atmosphere by rain and reach the sea surface in the form of precipitation. This process takes place in two stages, the first being *rainout*, where particles are removed in the rain cloud, possibly acting as nuclei for raindrop formation (see Fig. 1). The second stage is *washout*, where particles are scavenged from the atmosphere by falling raindrops. Scott [70] and Slinn [66] have given thorough reviews of washout factors and precipitation scavenging of particles.

There are certain problems and possible errors inherent in the measurements and calculations for wet removal of substances from the atmosphere. The use of washout factors to calculate wet deposition rates involves an implicit assumption that there is a linear relationship between the concentration of a substance in rain and air. This is difficult to prove. Possible differences exist in the particle removal efficiency as a function of chemical composition of the particles and rain-droplet size. Concentrations of substances in rain represent the integrated removal in the atmospheric column through which the rain is falling; however, air concentration is measured only at sampling tower level near the ground. Substances with different vertical concentration profiles but similar surface air concentrations may have different washout factors. Collections in three different terrestrial areas [71] demonstrated that the washout factor increased with increasing particle size at all three locations. Any relationship developed in urban or other continental regions between washout factors and particle size distribution, however, may not apply over open-ocean areas.

The enrichment factors for rain and atmospheric particles at Enewetak Atoll, calculated relative to average crustal rock, are given for different elements during

Table 5. Enrichment factors for rain and atmospheric particles calculated relative to average crustal rock[a] (From [63])

Element	Dry season		Wet season	
	Aerosol particles[b]	Rain	Aerosol particles[b]	Rain
Sc	0.78	0.26	0.63	0.65
Fe	0.89	0.68	0.81	0.71
Mn	0.86	0.96	1.1	0.69
Co	0.89	2.6	0.99	
Th	1.9	1.7	1.6	4.4
V	1.3	7.4	5.5	6.8
Cu	1.9	24	6.7	18
Zn	3.6	190	33	45
Pb	11	33	110	
Ag	83	420	1,200	
Cd	75	1,300	180	790
Se	3,700	1,900	48,000	64,000
I	10,000	46,000	73,000	230,000

[a] Data for average crustal rock from Taylor [27]
[b] Duce et al. [62]

Table 6. Estimates of the percentage of the trace element wet deposition associated with recycled sea spray (From [63]).

Element	Number of samples	Percentage recycled
Zn	9	15 ± 13
Pb	1	30
V	12	33 ± 32
Cu	10	48 ± 52
Se	10	17 ± 15
Ag	1	24

the dry and wet seasons in Table 5. Concentrations of Sc, Fe, Mn, Co, and Th in rain were dominated by crustal sources. On the other hand, enrichment factors for other elements (V, Cu, Zn, Pb, Ag, Cd, Se, and I) showed that non-crustal sources affected their concentrations in both aerosol particles and in rain [63].

Recent studies have demonstrated that a substantial proportion of the wet deposition arises from recycled sea spray [63]. Estimates derived from these studies suggest that Zn and Se are recycled to the extent of 15% of the total wet deposition (Table 6). The estimates ranged from 30% to nearly 50% for recycling of Pb, V, Cu, and Ag. Large uncertainties are associated with these estimates, partly because the ratios of the trace elements to Na in the rain samples varied during showers as well as between events, and partly because the estimates did not take into account the considerable variability [32] of enrichment factors for aerosol particles.

Gaseous Substances

Gaseous substances consist of true gases as well as liquids and solids in the vapour form. They behave differently than particles in the atmosphere and are subject to different processes. Photochemical processes can modify them and lead to gas-to-particle conversion. Particles formed in this way may be removed from the atmosphere by rainout and washout to reach the sea in wet precipitation.

Gases and vapours that are not dispersed in and mixed with the atmosphere obey the same basic laws of physics that control solid substances. Gases and vapours heavier than air settle while those that are lighter than air rise. When they come into contact with the sea surface, their entry into seawater is determined by a number of factors [72–74].

In a two-layer boundary system (see Fig. 2) at the air-water interface, the flux, F, through each boundary layer is given by:

$$F = k\varDelta C, \tag{2}$$

where $\varDelta C$ is the concentration difference across the particular layer and k is the corresponding exchange coefficient or transfer velocity. The coefficient, k, depends on many factors, including the degree of mixing of the water and air and the chemical reactivity of the gas. Its reciprocal is often called the resistance, r, which is a measure of the resistance to the transfer of the gas between air und water. The total resistance, R_t, expressed either on an air phase ($1/K_a$) or liquid phase ($1/K_w$) basis, depends on the exchange constants of individual phases and the value of the Henry's Law constant, H, for the gas concerned. Thus,

$$1/K_w = 1/k_w + 1/Hk_a, \tag{3}$$

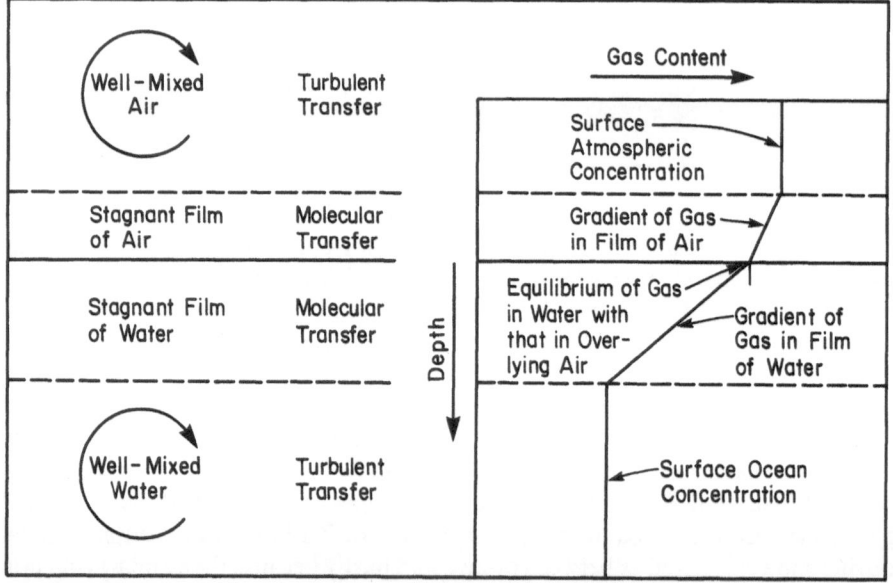

Fig. 2. Conceptual model of exchange of gases between the atmosphere and the ocean (Modified from Broecker and Peng [101])

For convenience in use of Eq. (3), $1/k_w$ may be written as r_w and $1/Hk_a$ as r_a, giving:

$$R_t(w) = r_w + r_a. \tag{4}$$

A relationship such as equation (4) can be used to ascertain the relative importance of either the air or liquid resistance for the exchange of any particular gas. Most laboratory and field studies have measured only the total resistance, because individual resistances are too difficult to measure. For all practical purposes, however, the total measured resistance is numerically equivalent to the resistance of the rate-controlling phase, inasmuch as the resistance of either one of the phases tends to predominate for most of the common atmospheric gases.

The phase in which the resistance to transfer of a gas between the atmosphere and the sea predominates is determined by the solubility of the gas in and reactivity with seawater. The air-sea exchange of relatively insoluble and unreactive gases, such as O_2, N_2, CO_2, CH_4, N_2O, and noble gases, is controlled by the resistance in the liquid phase. Those gases that are soluble in and possibly reactive with seawater, e.g., SO_2, SO_3, NH_3, NO_2, HCl, and HF, are controlled in their transfer from the atmosphere to the sea by the resistance in the gas-phase laminar layer.

Sea-to-Atmosphere Transfer

Normal diffusive and advective processes convey material through the water column to the sea surface. Gas bubbles may also transfer substances from subsurface seawater to the sea surface. Organisms, particularly plankton in their diurnal migrations, carry substances to the surface.

Various mechanisms for sea-to-air transfer of substances have been proposed. Bubble bursting appears to have the greatest support as the main mechanism for transferring substances from the sea to the atmosphere [75–78], but such processes as wind shearing of wave crests to produce spray [79] and rain-drop splashing have been regarded by various investigators as potentially important. The principal processes of sea-to-air transfer of substances are illustrated in Fig. 1. It has been suggested that the incidence of whitecaps in the oceans is a good measure of the transfer of substances from the oceans to the atmosphere. The global coverage of the oceans with whitecaps at any one time has been estimated to be 1 % [78], although some investigators [e.g., 79] argue that this figure should be smaller.

Particulate Matter

Blanchard and Syzdek [80] have demonstrated in a number of experiments the water-to-air transfer of bacteria and bacterial enrichment in drops from bursting bubbles. Similar studies demonstrated the transfer of virsus from water to air [81, 82]. Although these researches were designed mainly to investigate processes of enrichment of microorganisms on bubbles as they rise through the water column, they demonstrated the possibility of bacteria and viruses being transmitted from sewage-polluted seawater to the atmosphere and infecting terrestrial animals including man. Such a possibility has been borne out in the infamous Legionnaires' disease in which the bacterium *Legionella pneumophila* infected its victims through atmospheric transmission from contaminated water in cooling towers of air conditioning systems [83]. Blanchard [78] has thoroughly reviewed the production, distribution and bacterial enrichment of the sea-salt aerosol. Bubble bursting has been suggested [84, 75, 77] as the dominant mechanism for producing sea-derived aerosols. Blanchard [78] differentiates film drops from jet drops in a bursting air bubble, where film drops arise from the bursting bubble film while jet drops are created by the jet of water that rises rapidly from the bottom of the collapsing bubble cavity. He notes that film drops may be more important in the particle flux to the atmosphere than jet drops, although assumptions regarding film-drop production used in arriving at this conclusion have to be verified.

Gaseous Substances

Gases transfer from seawater to the atmosphere through a number of processes. When seawater is supersaturated with respect to a particular gas, bubbles of that gas are formed and normally rise to the sea surface. Larger bubbles usually break as they emerge from the seawater, releasing their gaseous contents to the atmosphere. Small bubbles will break when the sea surface is agitated by wave action, thus releasing gas to the atmosphere. Gases may also diffuse directly to the atmosphere from the sea surface.

Hydrophobic substances that form a film on the sea surface, such as petroleum hydrocarbons, will evaporate and in this way will contribute gaseous material to the atmosphere. Other substances, such as halogenated hydrocarbons and metals, present in a sea-surface film may also volatilize into the atmosphere, even though this form of transfer of normally solid substances from the sea-surface to the atmosphere is extremely slow.

Sea-Surface Microlayer

A thorough review of the sea-surface microlayer is given in the chapter by L. W. Lion (The Surface of the Ocean) [85] and the microlayer is discussed in another chapter by P. J. Wangerky (Organic Material in Sea Water) [10], relative to its role in connection with the dissolved and particulate organic matter in the sea. The microlayer is noted here only to the barest extent that is essential for a full discussion of air-sea exchange of pollutants.

The thin film at the surface of the sea, known as the microlayer, creates a form of discontinuity in the water column, where there is a temporary holdup of substances moving vertically (see Fig. 2). It is generally regarded as about 100 μm thick, but no set thickness defines the microlayer. Rather, the thickness of the microlayer is defined operationally by the technique used to sample it [86–88], and may range from 0.001 μm to 1000 μm. This is a drawback to quantitative determinations of substances in a microlayer of standardized thickness that would allow comparisons from one area to another.

A great deal has been written about the unique physical, chemical, and biological characteristics of the sea-surface microlayer [72, 87, 89–91]. It is the site of considerable biological activity [92–94]. Neuston organisms are enriched in the microlayer, compared to that in the seawater only a few cm below, by factors of 10^2–10^4 for bacterioneuston, 1–10^2 for phytoneuston and 1–10 for zooneuston [87]. Concern has been expressed by some investigators [e.g., 95] that high concentrations of metals, especially lead, in the microlayer can have a negative impact on phytoneuston productivity.

Substances entering the microlayer from the water column or from the atmosphere are temporarily held there. The residence time of substances in the microlayer may range from a few seconds to a day, depending on the nature of the substance and the wave activity. The estimated residence time for water-wettable particulate trace metals is about 2 seconds. When trace particulates are stabilized with less wettable organic coatings, the particles have estimated residence times in a 50 µm thick microlayer of from 1 to 30 minutes [96]. It was postulated by Pattenden et al. [97] that the residence time for most heavy metals in a microlayer of 1 µm thickness is 5 to 20 minutes. Hardy et al. [98] gave the mean residence time

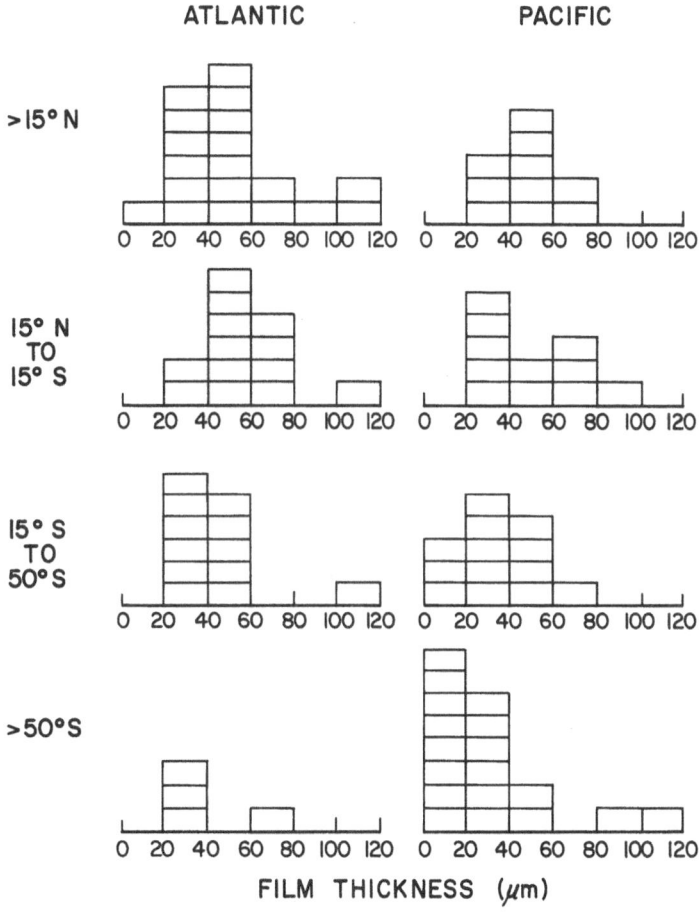

Fig. 3. Histograms of stagnant film-thickness estimates based on radon measurements in the surface waters at different latitudes of the Atlantic and Pacific oceans (From Broecker and Peng [101])

of metals in a microlayer of 50 μm thickness, based on calculations using a model and relevant data, as 3.5–15 hours in a calm sea, increasing through the series Ag, Zn, Mn, Pb, Cu, and Ni; but with a 4 m s^{-1} wind, the residence time varied from 1.5 to 8.5 hr, with Pb, Zn, Cu, Ni, Ag, and Mn giving 2.0, 1.5, 8.5, 2.4, 2.6, and 2.6 hr, respectively.

Various physical processes influence the concentrations of trace metals and organic carbon in the surface microlayer [99], including wind mixing and effects of rain [100]. Clearly, calm stagnant conditions enhance the stability of the surface microlayer. Broecker and Peng [101] have estimated the "stagnant film-thickness" at various latitudes in the Atlantic and Pacific oceans, using radon measurements. Histograms of stagnant film-thickness estimates based on these measurements are shown in Fig. 3 for different latitudes in the Atlantic and Pacific. The generally greater thickness of the film in the equatorial zone than in the temperate zones, and particularly in the Antarctic, is consistent with the fact that wind speeds are somewhat higher, on the average, over the Antarctic Ocean and somewhat lower over the equatorial ocean than over the temperate ocean. The authors suggested that the exchange rate for gases is somewhat lower than average in the equatorial zone (15°S to 15°N) and somewhat higher than average in the Antarctic zone (> 50°S).

Measurements of Substances for Air-Sea Exchange Determinations

In all measurements of substances for air-sea exchange determinations, two major points must be noted: (1) concentrations of substances of interest in both the atmosphere and the sea are usually extremely low, and analytical methods used must have a low limit of detection; and (2) extreme precautions must be taken to avoid contamination in both sampling and analysis. Some of the sampling techniques used for acquiring both atmospheric and seawater samples in the SEAREX program have been described [7].

Atmospheric

For ambient concentrations of substances in the atmosphere it is vital to take samples in locations well removed from local contamination. Continental stations, even when they are remote from urban environments, potentially suffer from local sources of contamination. Coastal stations that have been used for this purpose have been generally those where the prevailing winds are onshore and land-to-sea winds are comparatively infrequent. Site selection for atmospheric sampling generally, and for the Mediterranean specifically, has been described in GESAMP [7].

The choice for atmospheric samples without local contamination has been usually oceanic island stations. Bermuda has often been used for sampling the atmosphere over the northwest Atlantic. The SEAREX investigations have relied largely on island stations in the South Pacific such as Enewetak, American Samoa, Marshall, and Fanning islands. The Ninety Mile Beach site on the northwest tip of the North Island of New Zealand has also been utilized for SEAREX experiments.

In all cases, towers were erected to remove the atmospheric sampling point from ground-level contamination. Experience indicated that samplers should be located at least 10 m above ground level and as close as possible to the shoreline. Sampling has been controlled sometimes by wind direction, so that samples are taken only when the wind is coming from a particular desired direction. Utmost care has to be taken to avoid contamination of rain samples either by local spurious input or by sampling containers that have not been scrupulously cleaned.

It has been stressed [7] that because of the complexity of direct measurement of dry deposition, it should be avoided. Instead, dry-deposition fluxes can be estimated using aerosol-collecting techniques with size-separated aerosol collectors (cascade impactors) and relevant deposition models.

The experience in SEAREX suggests that rain sampling for wet deposition be done on an event basis rather than continuously. Improvements in instrumentation now allow such sampling to be done fully automatically. It has been recommended [7] that bulk air filtration sampling should be undertaken on a short time scale (12–24 hr). Such sampling might allow eastablishment of relationships between atmospheric contaminant concentrations and air mass movements. Shipboard air samples collected during oceanographic cruises can provide supplemental information.

Accuracy and precision are vital in analyses of the atmospheric substances obtained in dry and wet deposition and by air filtration. GESAMP [7] recommended that flameless atomic absorption spectrophotometry or anodic stripping voltammetry be used for trace metal analysis. Both of these techniques have been developed to a high degree of sensitivity, precision and accuracy. They also offer simplicity of operation for routine analyses. For organic substances, depending on the compounds chosen for analysis, the degree of specificity and sensitivity of analysis required and whether "finger printing" identification data are needed, GESAMP [7] indicated that a choice could be made from gas chromatography, high-pressure liquid chromatography, or gas chromatography – mass spectrometry.

Intercalibration of procedures and standardized techniques is essential if several laboratories are involved in analysis, in order to provide a measure of consistency in the data from one laboratory to another.

Marine

Sampling of seawater to avoid contamination has to be as meticulous as that for air samples. Ships are notorious for providing many sources of contamination, particularly from the exhausts of the propulsion and electricity-generating equipment. If seawater sampling is done from aboard an oceanographic vessel, it is essential to sample from the bow while the vessel is facing into the wind, and if possible, stemming a current. Launches that can be taken beyond the influence of an oceanographic vessel are sometimes used for contamination-free sampling of seawater.

While fuel residues and combustion products lead to organic contamination of seawater samples, oceanographic wire and sampling bottles can lead to contamination of water samples by metals. It is essential that the oceanographic wire and sampling bottles be coated with teflon or some other inert plastic material, or in the case of bottles, be constructed of polyvinyl chloride or some other suitable inert synthetic material.

For sampling of the sea-surface microlayer, various devices have been described [86–88], with advantages and disadvantages of each being given. A screen-type of sampler developed by Garrett [102] has probably received the widest acceptance. The Bubble Interfacial Microlayer Sampler (BIMS) [48–50] has been used in the North Atlantic, as described earlier, to determine enrichment factors for trace metals on bubble-derived sea-salt particles [32]. This device essentially samples the sea-surface microlayer by creating bubbles which burst at the sea-surface and then collecting for analysis the film and jet drops that are produced.

The same precautions are required to avoid contamination in analysis of substances collected in water as in air. Extraction procedures can lead to contamination. Therefore, reagents used must be absolutely pure, and handling has to be done in a way to avoid contamination. Techniques similar to those recommended for atmospheric substances may be used for analysis of metals and organic compounds in seawater. Again, intercalibration of procedures and standardized techniques is vital when several laboratories are involved in sampling and analysis of seawater for a particular cooperative project. Problems of inconsistency from one laboratory to another with difficult-to-measure elements, such as lead, are soon recognized in such an exercise [103, 104]. Suitable reference materials with confirmed concentrations of constituents are essential for such an intercalibration exercise. One seawater reference material for trace metals now available is NASS-1, produced by the National Research Council Canada, Ottawa [105]. The seawater used for this reference material came from an open-ocean sample, with a salinity of 35.07‰, collected at 1300 m depth in the North Atlantic, southeast of Bermuda.

Global Fluxes of Substances Between the Atmosphere and the Sea

There have been few direct flux measurements worldwide for substances exchanged between the atmosphere and the sea. Some measurements that can contribute to flux evaluations have been completed recently and continue to be made at oceanic island stations in the Pacific under the SEAREX program [32, 62, 63, 106]. Unfortunately, these data can hardly be extrapolated to a global scale.

Particulate Matter

Particulate matter entering the sea from the atmosphere consists mainly of windborne continental dust. Various estimates have been made on the global input to the oceans of aeolian dust. Nriagu [53] gave an estimate of the global production of windblown dust as 5×10^{14} g·yr^{-1}. Compared to that was an estimate of particulate production of volcanoes at 1×10^{13} g·yr^{-1} and an estimated total emission of particles from vegetation at 7.5×10^{13} g·yr^{-1}, a value considered to have a very large uncertainty according to Weisel et al. [32]. Chester

[107] gave data attributed to Goldberg [108] for atmospheric continental soil and rock particles at $1-5 \times 10^{14}$ g·yr^{-1}. It is uncertain how much of the global atmospheric dust reaches the sea, but it is likely no more than 70% and probably substantially less.

Recent studies in the SEAREX program have demonstrated that the seasonal variation in the amount of atmospheric dust at the Pacific oceanic island stations can be related to dust storm events in Asia [109]. The normalized temporal variations of the total dust flux, atmospheric dust concentration and rainfall rate for a station in Oahu is shown in Fig. 4. During the period from mid-January to mid-May the total dust fluxes were always above the annual average. On the other hand, the dust fluxes were very low during the summer, increasing slightly in the

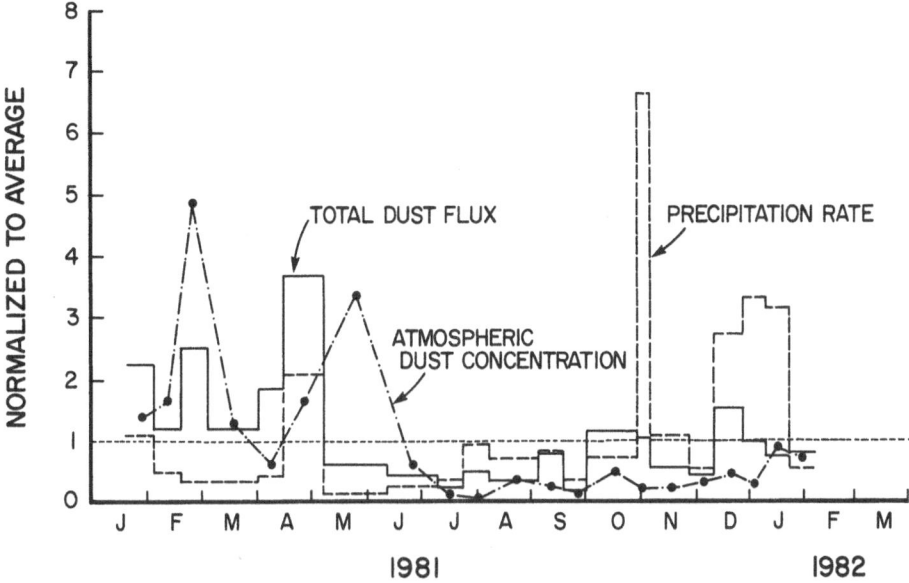

Fig. 4. Total dust flux, atmospheric dust concentration, and precipitation at Oahu, during 1981–1982 (From Uematsu et al. [109])

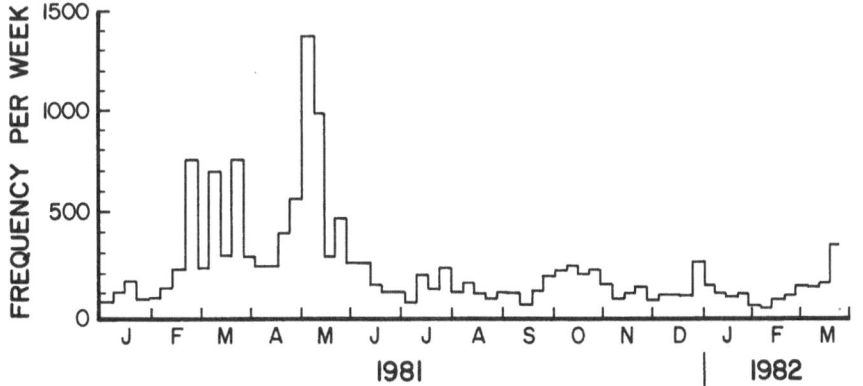

Fig. 5. Frequency of dust storm reports in Asia, during 1981–1982 (From Uematsu et al. [110])

fall. This pattern was similar to the seasonal variation in atmospheric dust concentration which was related to the dust storm activity in Asia (Fig. 5) [110].

Particulate matter in the atmosphere derived from the sea consists mainly of sea-salt aerosols. As noted in an earlier section, the currently accepted value for annual flux of sea-salt to the atmosphere is 3×10^{15} g, which translates into 1×10^{15} g of sodium transferred annually from the sea to the atmosphere [32]. From these values, fluxes of other inorganic trace constituents can be evaluated.

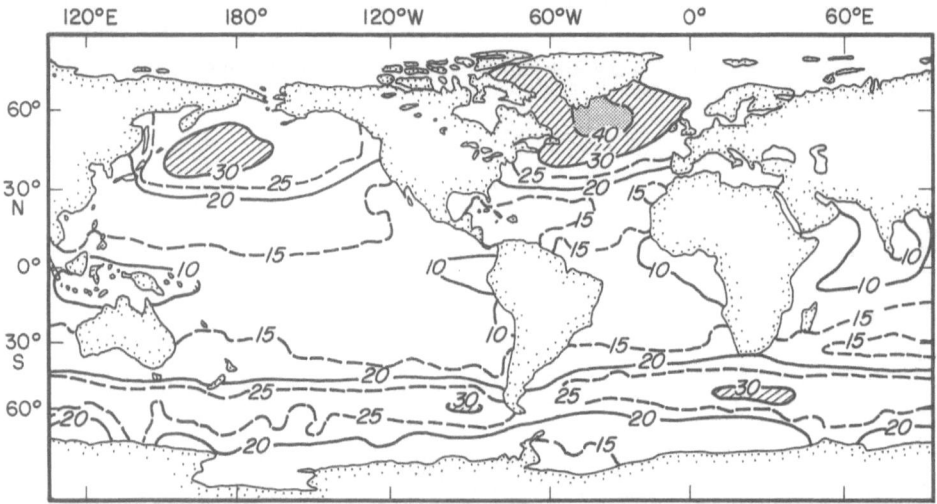

(A) DECEMBER – JANUARY – FEBRUARY

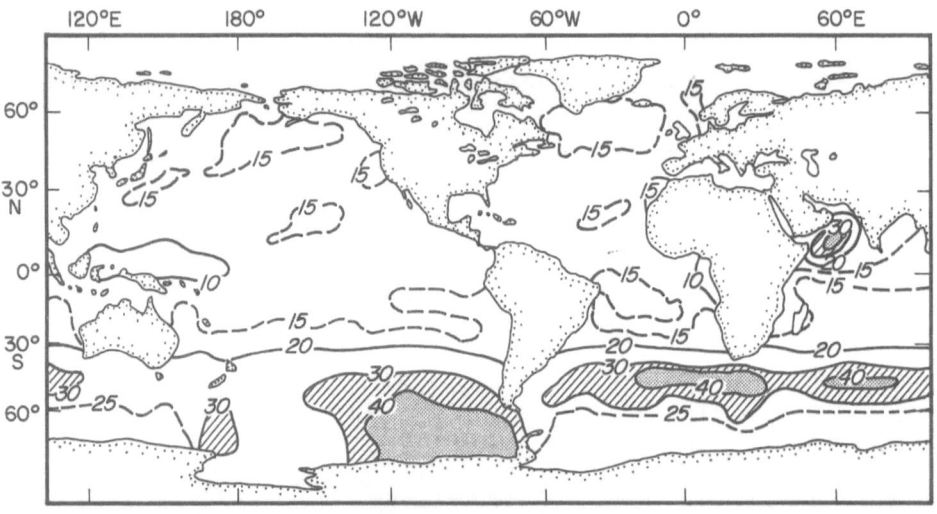

(B) JUNE – JULY – AUGUST

Fig. 6. Atmospheric seal-salt concentration ($\mu g \cdot m^{-3}$) distributions, during (A) the boreal winter (DJF) and (B) the boreal summer (JJA) (From Erickson et al. [112])

The seasonal distributions of sea-salt aerosols in the world oceans, calculated by Erickson et al. [111, 112], are shown in Fig. 6. To quantify the global annual production of atmospheric sea-salt, the authors used empirical relationships between wind speed and atmospheric concentrations of sea-salt 15 m above the sea surface. Using data on global surface winds, they developed seasonal maps of sea-salt concentration distributions over the world oceans. Figure 6(A) shows the atmospheric sea-salt concentration distributions during the boreal winter (Dec.-Jan.-Feb.), while Fig. 6(B) shows these distributions during the boreal summer (Jun.-Jul.-Aug.). The authors noted that the continuous strong winds at high latitudes in the southern hemisphere result in a high and relatively constant atmospheric sea-salt concentration ranging from 20 to 40 $\mu g\ m^{-3}$ from the austral summer to winter. On the other hand, winds in the high latitudes of the northern hemisphere exhibit quite a large seasonal variation and result in a greater than a factor of 3 difference in atmospheric sea-salt concentrations between the boreal summer and winter. Using these seasonal distributions and the residence time for a sea-salt aerosol one could make a further rough estimate of global annual production of sea-salt aerosols.

Trace Elements

The concentrations of trace elements in precipitation of different regions of the world are shown in Table 7, where geometric means of volume-weighted data are

Table 7. Comparison of the concentrations ($\mu g\ kg^{-1}$) of trace elements in precipitation from different regions (From [63])

Element	Enewetak[a] [63]			Bay of Bengal [113]	Collafirth [13]	Bermuda [114]
	Dry season	Wet season	Annual			
Na	1,700	1,000	1,100	14,000	22,000	3,400
Mg	290	150	170	1,600		490
K	49	38	39	1,400		160
Ca	78	48	50	1,700		310
Cl	3,100	1,800	2,000	25,000	46,000	6,800
Br	11	6.5	7.1		360	
Al	4.3	1.9	2.1		230	
Fe	2.1	0.93	1.0	30	135	4.8
Mn	0.043	0.010	0.012		5	0.27
Sc	0.00032	0.00022	0.00023	0.016	0.021	
Th	0.00083	0.00093	0.00091	0.12	0.07	
Co	0.0021			0.95	0.190	
I	1.3	1.2	1.2			
V	0.068	0.014	0.018		8	
Zn	0.43	0.043	0.052	100	26	1.15
Cd	0.0085	0.0018	0.0021			0.06
Cu	0.071	0.011	0.013		23	0.66
Pb	0.085	0.032	0.035		11	0.77
Ag	0.0086	0.0054	0.0056			
Se	0.011	0.023	0.021		0.40	

[a] Geometric means of the volume-weighted data

Table 8. Concentrations of Pb and Cd in the atmosphere and precipitation, and total annual deposition of these elements in different regions of the world oceans (From [7])

	Pb	Cd	References
Atmosphere		$ng\,m^{-3}$	
Samoa Area,	0.02	0.002	Duce, unpublished data
Tropical South Pacific			Patterson, unpublished data
Enewetak,			
Tropical North Pacific	0.12	0.003	[62]
Hawaii	2	0.02	[115, 116]
North Atlantic	10	0.13	[117]
Bermuda Area	3	0.2	[118]
Baltic Sea	10–60	0.1–0.5	[119]
North Sea	20–200	0.5–2.5	[120]
			[121, 122]
Mediterranean Sea	10–60	0.4–2.1	[14, 123]
			Buat-Ménard et al., unpublished data
Precipitation		$\mu g \cdot l^{-1}$	
Samoa	0.007	–	Duce et al., unpublished data [124]
Enewetak	0.023	0.0041	[63, 124]
Bermuda	0.77	0.006	[114]
Baltic Sea	10–30	0.3	[119]
North Sea	10–35	0.3–1.2	[120]
Mediterranean Sea	6–12	–	Buat-Ménard et al., unpublished data
Total deposition		$ng \cdot cm^{-2} \cdot yr^{-1}$	
Enewetak	7	0.35	[63]
North Atlantic	310	5	[117]
Baltic Sea	400–1750	13–20	[119]
North Sea	700–2600	20–85	[120]
Mediterranean Sea	300–1800	10–50	Calculated from [121, 122, 14] and Buat-Ménard et al., unpublished data

presented. Concentrations of Na, Mg, K, Ca, Cl, and Br in Enewetak rain samples were similar to those for precipitation in other regions of the world [63]. Higher concentrations of sea-salt elements occurred during the dry season which were probably due to higher wind speeds in the dry season. Concentrations of Pb and Cd in the atmosphere and precipitation, and flux data for these elements for different marine regions, are given in Table 8. It is obvious that there are large increases in both concentrations and fluxes from oceanic regions to areas where man-made inputs predominate.

The estimated mean fluxes of trace elements from the atmosphere to the sea-surface in various coastal and oceanic regions of the world are given in Table 9. It should be noted that the flux estimates for the different regions were based on a variety of assumptions which were not necessarily the same among the various investigators. This may account in part for the large variations in fluxes among the different regions for such trace metals as Cu and Zn.

Table 9. Estimated mean fluxes of trace metals from the atmosphere to the sea surface $(ng\,cm^{-2}\,yr^{-1})^a$ (From [6])

Element	New York Bight [12]	North Sea [13]	Western Mediterranean [14]	South Atlantic Bight [11]	Bermuda [118]	Tropical North Atlantic [125]	Tropical North Pacific [126]
Al	6,000	30,000	5,000	2,900	3,900	5,000	1,900
Sc	–	5	1	–	0.6	1.1	0.4
V	–	480	–	–	5	1.7	2.8
Cr	–	210	49	–	9	14	6
Mn	–	920	–	60	45	70	18
Fe	5,700	25,500	5,100	5,900	3,000	3,200	1,300
Co	–	39	3.5	–	1.2	2.7	0.6
Ni[b]	–	260	–	390	3	20	–
Cu[b]	–	1,300	96	220	30	25	2
Zn[b]	1,400	8,950	1,080	750	75	130	13
As[b]	–	280	54	45	3	–	–
Se[b]	–	22	48	–	3	14	4.5
Ag[b]	–	–	3	–	–	0.9	0.2
Cd[b]	30	43	13	9	4.5	5	0.5
Sb[b]	–	58	48	–	1.0	3.5	0.1
Au[b]	–	–	0.05	–	–	0.1	–
Hg[b]	–	–	5	24	–	2.1	–
Pb[b]	3,900	2,650	1,050	660	100	310	7
Th	–	4	1.2	–	–	0.9	0.9

[a] Flux estimates are based upon a variety of assumptions, not necessarily the same among the listed investigators
[b] Designates elements generally enriched in marine aerosols

Organic Constituents

Organic substances are emitted to the atmosphere by a vast array of natural and anthropogenic sources. They comprise an enormous number of organic compounds each of which has its individual physical and chemical characteristics and unique behaviour in the environment. Little is known about the global sources, distribution and fluxes of this organic material. Attention has been focussed on two particular groups of organic compounds because of their environmental and ecological effects: halogenated hydrocarbons; and petroleum hydrocarbons. The effects of petroleum hydrocarbons are obvious if one has seen even a minor oil spill. The effects of halogenated hydrocarbons are less obvious, but serious concern has been expressed by both the scientific community and the general public about their potentially adverse ecological and human health effects.

Halogenated Hydrocarbons

The halogenated hydrocarbons can be conveniently subdivided into two groups: the high-molecular-weight chlorinated hydrocarbons; and low-molecular-weight halogenated hydrocarbons or halocarbons. The first group consists of the well known organochlorine pesticides, e.g., DDT, and the industrial organochlorine chemicals, such as the PCBs (polychlorinated biphenyls). The second group, the halocarbons, stem from the use of these gaseous substances in pressurized aerosol dispensers and in refrigeration. Freon-11 (trichlorofluoromethane, CCl_3F) is a prime example. Unlike the high-molecular-weight halogenated hydrocarbons, which are virtually all man-made, some of the halocarbons, particularly the monohalomethanes, have natural sources in the oceans.

High-Molecular-Weight Chlorinated Hydrocarbons

The ecological effects of organochlorines first manifested themselves in eggshell thinning in birds leading to reproductive failure through egg breakage from high concentrations of DDT and its metabolites, especially DDE. The PCBs have aroused much public worry, because of their perceived carcinogenic properties, but except for chloracne, a skin condition in humans, and an association of high concentrations of these compounds with reproductive failure in marine mammals and toxicity to minks, the ecological and human health impacts of PCBs have not been too well established [127]. What has created much interest in the scientific community is that these man-made halogenated hydrocarbons have been found in marine animals in areas, e.g., Antarctica, far removed from sites of application and release of these compounds [128, 8, 9]. Clearly, these substances have been transported virtually around the globe by tropospheric winds. As many developed countries have banned the use of DDT and other ecologically harmful pesticides, their utilization has shifted to developing countries in the tropics and the southern hemisphere, where DDT spraying is still the most effective means of controlling the mosquito that spreads malaria.

Concentrations of DDT were reported off the California coast, ranging from 2 to 6 $ng \cdot l^{-1}$ in 1971 [129], 0.2 to 1.8 $ng \cdot l^{-1}$ in 1973–1974 [130, 131] and 0.05 to 0.8 $ng \cdot l^{-1}$ in 1975 [132]. Subsurface waters in the North Atlantic gave DDT concentrations below the detection limit of 0.1 $ng \cdot l^{-1}$ [8, 133], and in the North Central Pacific 0.03 $ng \cdot l^{-1}$ [134]. DDT has been measured somewhat more reliably in surface films of the open ocean, because of higher concentrations there

Table 10. Concentrations in the atmosphere ($ng\,m^{-3}$) of organic compounds in different marine regions (From [7])

Region	PCB	n-Alkanes (Vapour)	n-Alkanes (Particulate)
Mediterranean	0.04–0.3 [139]	65–147[a] [143]	10.8–43.7 [143]
North Sea	0.96 [140]	–	–
Central Pacific	0.19–0.32 [141]	–	–
North Pacific Trades	0.049 [142]	2.6[b] [144]	0.044[c] [144]
South Pacific Trades	0.012 [142]	1.4–3.9[d] [145]	–
West Ireland	–	253[e] [146]	3.3[d] [146]
Equatorial Atlantic	–	30–281[a] [147]	1.5–14[a] [148]
Tropical North Atlantic	0.21–0.65 [8]	–	4–50[a] [148]
North Atlantic	–	66[b] [144]	3.3[d] [144]

[a] C_{14}–?   [b] C_{13}–C_{30} [c] C_{21}–C_{30} [d] C_{15}–C_{28} [e] C_{10}–C_{28}

[8]. The global flux of DDT from the atmosphere to the oceans following its peak use in the northern hemisphere was estimated at $2 \times 10^8 \, \mathrm{g \cdot yr^{-1}}$ [135, 136].

PCB concentrations in California coastal waters were reported at $2.5 \, \mathrm{ng \cdot l^{-1}}$ in 1971 [134] and <0.1 to $0.7 \, \mathrm{ng \cdot l^{-1}}$ in 1975 [132]. In the North Atlantic, they ranged from 5 to $40 \, \mathrm{ng \cdot l^{-1}}$ in 1971 [135, 137]. PCB concentrations in the North Atlantic were found to decrease from $27–41 \, \mathrm{ng \cdot l^{-1}}$ in 1972 to $0.8–2.0 \, \mathrm{ng \cdot l^{-1}}$ in 1973 [138, 133]. The decrease was attributed to the combined effects of scavenging of the PCBs by sinking particles and declining inputs arising from a ban on use of PCBs [133]. Recent concentrations of PCBs in the atmosphere of different marine regions are given in Table 10. The flux of PCBs from the atmosphere to the oceans following their peak use in the northern hemisphere was estimated at $2 \times 10^9 \, \mathrm{g \cdot yr^{-1}}$ [149, 136].

Low-Molecular-Weight Halogenated Hydrocarbons

The halocarbons have been a source of environmental concern not so much because of direct effects on the atmospheric or aquatic environments but because of their impact on the stratospheric ozone layer. They are now known to cause erosion of the ozone layer that is so vital for protection of the earth's life against damaging ultraviolet rays [150, 151]. There does not appear to be a great flux of these halocarbons from the atmosphere to the oceans. Estimates for trichloro-fluoromethane (Freon-11, CCl_3F) show that the oceans are a sink for 0.5–1.0% of the atmospheric burden of this halocarbon, which is comparable to the stratospheric sink (1%) [152]. From measurements in the Atlantic [149] it was observed that CCl_3F was confined primarily to surface waters at that time, and that its concentration decreased rapidly with depth. Using measurements by Lovelock [153], Liss and Slater [74] calculated that $5.4 \times 10^9 \, \mathrm{g \cdot yr^{-1}}$ of Freon-11 are removed from the atmosphere by the oceans. This represented about 2% of the total world production of F-11 at that time. The chlorofluorocarbons appear to have long tropospheric lifetimes because of their low reactivity in the troposphere and their low solubility in water. Therefore, it could be expected that their long tropospheric residence time would allow the chlorofluorocarbons to undergo relatively complete tropospheric mixing between the northern and southern hemispheres.

Fluxes of some of the halocarbons are given in Table 11.

Petroleum Hydrocarbons

The petroleum hydrocarbons, derived as fossil fuel from the earth's crust, are a complex mixture of different organic constituents which vary from one oil field to another. When fractionated in refineries to produce different products for different uses the composition is further changed. Thus the introduction into the atmosphere of petroleum hydrocarbons will depend on the type of product, volatility of individual constituents and products of combustion, if they are used as a thermal energy source. To further complicate matters, there is a wide variety of biogenically produced hydrocarbons in the oceans which are difficult to separate from the petroleum hydrocarbons.

Numerous studies have been done on the sources and impact of petroleum hydrocarbons on the marine environment [154, 155]. It was estimated by the Workshop on Inputs, Fates, and Effects of Petroleum in the Marine Environment, held in Airlie, Virginia, in May 1973 [154], that the annual global input of petroleum hydrocarbons from the atmosphere to the oceans was 0.6×10^{12} g out of a total annual input of 6.113×10^{12} g. The atmospheric input has not changed in more recent evaluations although the total estimated annual input of petroleum hydrocarbons to the marine environment had substantially decreased by 1978 to 4.95×10^{12} g [156]. The 1982 U.S. National Academy of Sciences Petroleum in the Marine Environment Symposium concluded that 3.2×10^{12} g of petroleum-derived hydrocarbons enter the oceans each year, of which 0.3×10^{12} g comes from the atmosphere [157].

It has been estimated that the total northern hemisphere tropospheric burden of gaseous hydrocarbons is 91×10^{12} g, with 11×10^{12} g over the oceans and 80×10^{12} over land, and the annual anthropogenic and natural productions are 45×10^{12} g and 150×10^{12} g, respectively [158, 159]. These investigators estimated that the global annual man-made, "long-lived" gaseous hydrocarbon production is about 50×10^{12} g of which only 5×10^{12} g is produced in the southern hemisphere. They estimated a mean tropospheric residence time of 0.5 to 2.0 years for all hydrocarbons in the northern hemisphere that are not attached, or converted, to particles.

The annual input of nonmethane petroleum hydrocarbons to the atmosphere from marine transportation sources, offshore production facilities and natural oil seeps has been estimated at 1.35×10^{12} g [160]. An air-to-sea flux of 1.4×10^{12} g was determined by Duce [161], but the composition of this flux was quite different from the foregoing sea-to-air flux, in that the latter [161] consisted of particulate matter formed from the heavier organic constituents of petroleum and their reaction products, which were carried to the sea by dry fallout and precipitation processes.

More recent studies [144] have estimated that the total annual input of n-alkanes ($n\text{-}C_{10}$ to $n\text{-}C_{30}$) from the atmosphere to the oceans is 0.04×10^{12} g to 0.4×10^{12} g. Thus, based on the latest estimate [157] of 3.2×10^{12} g of petroleum-derived hydrocarbons entering the ocean each year, a maximum of 10–15% of the total petroleum input is entering the oceans from atmospheric sources. The n-alkanes constitute only one class of organic compounds in the marine atmosphere. In some open-ocean areas, fatty acids, alcohols, and wax esters have been found in atmospheric concentrations equal to or greater than the n-alkanes [162, 163].

Concentrations of n-alkanes in vapour and particulate form in the atmosphere of different marine regions are given in Table 10.

Gases

As noted in an earlier section, gas exchange between the atmosphere and the sea cannot be easily measured directly at present. Various models, however, can be used to measure the fluxes of gases between the atmosphere and the sea indirectly [72–74]. The net global fluxes across the air-sea interface have been evaluated for a number of gases by Liss [73], and these are reproduced in Table 11. There are

Table 11. Net global gas fluxes across the air-sea interface (Modified from [73])

Element	Compound	Controlling resistance[a]	Global air-sea flux		Method[d]	Reference
			Direction[b]	Magnitude[c]		
H	H_2	r_w	+	10^{12}	o	[164]
C	HCHO	$r_a \sim r_w$	−	10^{13}	o	[165]
	CH_4	r_w	+	10^{12}–10^{13}	o	[166]
	C_2H_6	r_w	+	10^{12}	o	[167]
	C_3H_8	r_w	+	10^{12}	o	[167]
	C_2H_4	r_w	+	10^{12}	o	[167]
	C_3H_6	r_w	+	10^{12}	o	[167]
	CO	r_w	+	10^{14}	o	[168]
	CO_2 (man made)	r_w	−	2×10^{15} (as C)	×	[169]
N	N_2O	r_w	+	6×10^{12}	o	[170, 171]
	NO	r_w	+(?)			[172]
	NO_2	?	+(?)			[173]
	NH_3	r_a	+(?)	?	o	[174]
O	O_3	r_w	−	6×10^{14}	o	[73]
S	Total volatile S		+	34–170×10^{12} (as S)	×	[175, 176]
	H_2S	r_w	+	15×10^{12}	×	[177]
	$(CH_3)_2S$	r_w	+	30–50×10^{12}	o	[178, 179]
	CS_2	r_w	+	0.3×10^{12}	o	[73]
	COS	r_w	+	0.8×10^{12}	o	[180]
	SO_2	r_a	−	5×10^{12}	o	[73]
Cl	CH_3Cl	r_w	+	3–8×10^{12}	o	[181, 182]
	CCl_4	r_w	−	10^{10}	o	[74]
	CCl_4	r_w	=	~ 0	o	[183]
	CCl_3F	r_w	−	5×10^9	o	[74]
	CCl_3F	r_w	=	~ 0	o	[184]
	$CHCl_3$	r_w	+	7×10^{11}	o	Rasmussen (pers. comm., 1982)
I	CH_3I	r_w	+	3×10^{11}	o	[74]
	CH_3I	r_w	+	13×10^{11}	o	[185]

[a] r_a = resistance in air; r_w = resistance in water
[b] +, sea→air; −, air→sea; =, no net flux
[c] Units are g (of the compound) yr^{-1}, except as indicated
[d] o calculated by $k\Delta C$ techniques (see text); × calculated by other technique described by Liss [73]

some more recent data, but these, e.g., that for methane given by Khalil and Rasmussen [186], are often of the same order of magnitude. Note the comparatively large number of gases that have a net flux from the sea to the atmosphere. These are generally considered to have natural sources in the marine environment.

Gases entering the sea from the atmosphere can be impeded from penetrating into deeper water by vertical density stratification developed by surface heating or freshwater runoff. This has been observed for vertical penetration of oxygen into seawater, for example, overlain by a layer of freshwater in the laboratory [187]. Many examples can be found for impedance of vertical aeration by salinity stratification due to freshwater inflow in coastal marine inlets [e.g., 188]. This impedance of transfer of gases from the sea-surface layer to deep water by vertical density stratification retards the entry of gases from the atmosphere to the sea-surface layer by reducing the partial pressure difference of the gas in the atmosphere and the sea-surface layer. Oil films, on the other hand, have a rather minimal effect on the transfer of oxygen into seawater, mainly because oil films are seldom intact over large areas for an extended period of time [189]. This probably applies to the effect of oil films on the transfer of other gases from the atmosphere to the sea as well.

In the following, some details are provided for the global balance and air-sea exchange only for carbon dioxide and methane. Carbon dioxide, of course, is a well recognized "greenhouse" gas, while methane is potentially important also as contributing to the greenhouse effect, and thereby, increasing global atmospheric temperatures.

Carbon Dioxide

As noted in the Introduction, the oceans play a vital role in buffering the increasing concentrations of carbon dioxide in the atmosphere arising from fossil fuel combustion. But there is a great deal of uncertainty about the extent of this buffering effect. It is generally accepted that there is a net flux at present of anthropogenic CO_2 from the atmosphere to the sea, amounting to about 2×10^{15} g·yr^{-1} (see Table 11) [169, 17]. The current amount of CO_2 entering the atmosphere annually from fossil-fuel burning is estimated to be 5×10^{15} g [16, 17]. The additional CO_2 in the atmosphere amounts to about 2.5×10^{15} gC yr^{-1}, which leads to the well documented increase of about 1.0 ppmv (parts per million by volume) per year since good measurements commenced at the Mauna Loa Observatory, Hawaii, in 1958, when the CO_2 concentration was 315 ppmv [16]. One of the uncertainties in the global carbon balance is in the amount of anthropogenic CO_2 that is retained in the atmosphere and contributes to the observed annual increase in concentration. This ranges from 30% to 60%, depending on the technique used for estimation, but is generally accepted at about 50% [190, 191].

Another major uncertainty in the global carbon balance is the biotic contribution of CO_2, due to deforestation and agriculture, to the atmosphere. Using three different sets of data for conversion of forest land to agriculture, Woodwell et al. [17] estimated for 1980 that between 1.8×10^{15} and 4.7×10^{15} g of carbon were released by the biota of which nearly 80% was due to deforestation, primarily in the tropics. Elliott et al. [16] examined statistically the three scenarios developed by Woodwell et al. [17], using the CO_2 concentrations measured at the Mauna Loa Observatory, Hawaii, for their analysis. They concluded that a net annual biotic addition as large as 0.7×10^{15} g is allowed by their analysis, but it could be equally possible to have a net loss to the biota of 0.2×10^{15} gC yr^{-1}. A major assumption in their analysis was that the year-to-year trend in the biotic contribution of CO_2 is not parallel to the trend in the fossil-fuel contribution of

CO_2. It is generally agreed that while the fossil-fuel contribution of CO_2 continues to increase the biotic contribution reached a peak around 1900, and then went into a decline, equalling the fossil-fuel contribution around 1960 [101]. It is difficult to justify from the present information that the net annual biotic contribution of CO_2 to the atmosphere is more than about 1×10^{15} gC.

The discrepancy between sources and sinks for man-made CO_2 ranges from 0.5 to 4.5×10^{15} gC yr^{-1}, depending on the net biospheric input. The oceans are the major identified sink for man-made CO_2, and to obtain a global carbon balance, either the oceanic sink is underestimated, or there are other as yet unidentified sinks. The median values (in 10^{15} g) for the global annual carbon balance equation proposed for 1980 by Woodwell et al. [17] appear to be applicable in 1986: Increment of atmospheric carbon (2.5) = Fossil fuel release (5.2) − Oceanic uptake of carbon (2.0) + Terrestrial carbon release (3.3) − Imbalance (4.0).

Methane

An n-alkane, methane is sometimes called "marsh gas" and is the simplest and most uniformly abundant hydrocarbon globally. It is associated with petroleum in fossil-fuel production and also has biogenic origins. Methane is the product of anaerobic decomposition of organic matter, and it has been estimated that 553 $\times 10^{12}$ g CH_4 are produced annually, of which 13×10^{12} g is the marine contribution [186]. Studies of CH_4 in bubbles of polar ice cores have shown that atmospheric concentrations of CH_4 several hundred to several thousand years ago were one-half to one-third of the CH_4 concentrations today [192, 193]. There is evidence that the concentrations of atmospheric CH_4 have been increasing along with the growth of human populations during the last century [194]. The concentration of CH_4 by mid-1982 was at about 1600 ppbv (parts ber billion by volume) at the Mauna Loa Observatory, Hawaii, and since January 1981, it had been increasing at the rate of 2.5 ± 1.2 ppbv per month [186].

Wetlands, enteric fermentation in cattle and other herbivores and waterlogged soils, such as rice paddy fields, producing 150, 120 and 95×10^{12} g CH_4 yr^{-1}, respectively. have been considered to be the main contributors of atmospheric CH_4 [186]. Other contributions are comparatively small (given in units of 10^{12} g · yr^{-1}): oceans, 13; freshwater lakes, 10; tundra, 12; biomass burn, 25; direct anthropogenic, 40; and other, 88. The "other" category includes the contribution of CH_4 from termites, a newly discovered source, which has been considered to be quite significant in recent studies [195, 196]. Tropical rain forests in particular, and natural land and water ecosystems in general, have been considered by other investigators [197] to be the largest sources of atmospheric CH_4. However, Khalil and Rasmussen [186] considered that there are large uncertainties in the source estimates of Sheppard et al. [197], and found it difficult to evaluate their final results as to their accuracy and validity.

The increase in anthropogenic CH_4 has been estimated to be 1.2–2% per year. There are about 4000×10^{12} g of CH_4 in the earth's atmosphere, suggesting that 50 to 80×10^{12} g more CH_4 are introduced into the atmosphere than removed annually. Accepting 500×10^{12} g-yr^{-1} of CH_4 as the steady-state contributions, one can calculate that the residual CH_4 amounts to 10–15%, a surprisingly large fraction, of the total [198]. This can be explained by the long life time (7–8 years) of CH_4 in the atmosphere, so that even a net increase of 10×10^{12} g of CH_4 annually over several decades can lead to such a large percentage increase.

Latitudinally, there is only about 7% more methane in the northern hemisphere than in the southern hemisphere. Because the oceans contribute annually such a small proportion (13×10^{12} g) of the total atmospheric methane, one must conclude that it must largely come from tropical rain forests in the southern hemisphere.

Methane is one of the radiatively active trace gases that can influence the radiative energy balance as easily as CO_2 [199]. It has been suggested by a number of investigators that the combination of CH_4 with other radiatively active trace gases could lead to a warming of the surface-tropospheric system, which may be of the same magnitude as that caused by increasing atmospheric CO_2 concentrations. Methane differs from CO_2, which is chemically relatively inert, in that it is quite reactive and plays a vital role in tropospheric chemistry and air quality. Through photochemical and other complex reactions, for example, increasing CH_4 concentration may lead to greater amounts of carbon monoxide and ozone. Increased CH_4 concentrations may ameliorate the depletion of the stratospheric ozone layer, and thus maintain the natural ozone screen for ultraviolet light, inasmuch as methane removes chlorine atoms from the stratosphere.

Methane reacts with the hydroxide ion free radical, OH, in the atmosphere, and this appears to be the main sink for CH_4 [186]. Thus the seasonal and regional variation of OH can be related to the variation in CH_4. The low winter and high summer concentrations of OH in part explain the seasonal cycle of CH_4 concentration, even though the atmospheric lifetime of CH_4 is estimated to be 7–8 years [186]. Depletion of OH radicals in the troposphere could result from increasing CO levels from industry, and as a result, the reduction of CH_4 by OH could be diminished. Changes of OH, CO, and CH_4 are coupled together over the long term [193], so that even a slow decline in OH radicals may enhance CH_4 concentrations. Early atmospheric measurements of OH [200] in the latitude range of 31° to 21° N and altitudes of 7 to 11.5 km gave values ranging from 3×10^6 to 8×10^6 radicals per cm^3 in October 1975. Later, budget considerations of $CHCl_3$ implied much smaller OH concentrations of $3 \times 10^5 \, cm^{-3}$ in the northern hemisphere and $7 \times 10^5 \, cm^{-3}$ in the southern hemisphere [201]. Atmospheric OH concentration is emerging as an important index of atmospheric pollution and the potential for removal of undesirable constituents.

Discussion

Determination of fluxes of substances between the atmosphere and the sea is fraught with many problems. Accurate measurement of concentrations in rain and dry deposition is a necessary first step. But this in itself is not sufficient. The total amount of rain in a given period, e.g., a year, is essential, and determination of this value has potential problems. The variability of rainfall amount and intensity with season can have an effect on the metal concentrations. Seasonal changes in wind flow patterns or particle production processes will affect atmospheric metal concentrations at a given marine locality.

Even if the foregoing problems are resolved, other significant problems remain. Recent studies [63] have shown that the gross deposition to the ocean is composed of a net input and a component associated with recycled sea spray. To evaluate the transport of materials from the continents to the oceans accurately one must be able to distinguish the net and recycled components [114, 115, 63]. Atmospheric sea-salt particles produced by bubbles bursting at the sea surface appear to contain many metals in concentrations considerably higher than would be expected on the basis of the metal-to-sodium ratio of near-surface water. Some fraction of these metals is apparently associated with surface-active organic material and is scavenged by rising bubbles and concentrated on the sea-salt particles produced from the sea-surface microlayer when the bubbles burst [32]. Therefore, the calculated net deposition of metals to the ocean will be anomalously high, if this fractionation is not taken into account. An extensive discussion of this problem was also given by Buat-Ménard [117].

One approach to solving this problem was to determine the metal/Na ratio on the largest atmospheric particles at Enewetak Atoll using only particles collected in the first stage of the cascade impactor [63]. Thus, particles were collected with radii greater than about 3.5 µm. Sea-salt particles are generally in this size range so that a good estimate of the metal/Na ratio was obtained for particles produced by bursting bubbles. There was good agreement between these ratios and the direct measurements of the metal/Na ratios on sea-salt particles collected using the Bubble Interfacial Microlayer Sampler in the North Atlantic [32]. The recycled component of any metal in the rain was then calculated, and for the metals Pb, V, Cu, and Ag it ranged from 15% to almost 50% (see Table 6). These values are comparable to the recycled component calculated by a similar approach at Bermuda [114], but the uncertainties in the estimations are large. The recycled component for Pb in Enewetak rain was calculated to be about 30% by Arimoto et al. [63], compared to 10% for the same atoll determined by Settle and Patterson [115] and 17% at Bermuda by Jickells et al. [114].

Recycled components calculated in the same way for dry deposition measurements at a surrogate surface ranged from 12 to 100%. The percentage of Pb recycled varied substantially between samples, with approximately 50 to 60% of the gross Pb deposition attributed to recycled material in two samples, and virtually 100% being considered due to sea spray in a third sample [71].

It has been proposed that more sophisticated techniques be developed for accurately determining this recycled fraction. Duce [71] suggested that future work should focus on two areas: (a) the use of suitable tracers (stable or radioactive) during field measurements; and (b) carefully designed *in situ* or laboratory experiments to improve our knowledge of metal/Na ratios as a function of sea-salt particle size.

Conclusions

1. There is evidence from various regional studies that the input to the sea of some contaminants from land via the atmosphere is of the same order of magnitude as riverine input.

2. Washout by rain (wet deposition) and dry fallout (dry deposition) are the main processes by which atmospheric particles and gases enter the sea.

3. Calculations of wet and dry deposition of inorganic and organic substances from the atmosphere have been made for a number of oceanic areas, particularly in the Pacific.

4. The global production of aeolian dust has been estimated at 5×10^{14} g · yr^{-1}. Studies in the Pacific have shown that pulses of high input of atmospheric dust to the oceans follow wind storm events in Asia.

5. Bubble bursting is the principal process by which sea-salt aerosols enter the atmsophere from the sea.

6. The annual global sea-salt aerosol production has been estimated at 3×10^{15} g. Sea-salt aerosol distributions over the world oceans have been seasonally mapped, based on mean surface wind speeds. Highest concentrations of over $40 \mu g \cdot m^{-3}$ occur during the boreal winter (DJF) over the North Atlantic, with a maximum at 50°–60° N, 20°–40° W, and during the boreal summer (JJA) over the South Atlantic and South Pacific from 40° to 80° S.

7. As much as 50% of certain metals reaching the sea surface from the atmosphere may be recycled to the atmosphere.

8. SEAREX studies have shown that the ocean is the prime source of atmospheric K and Mg. The ocean may also account for 5–20% of the global input of Cu, V and possibly Zn to the atmosphere. It is an insignificant contributor to the atmosphere of Al, Co, Fe, Mn, Pb, and Se. In remote parts of the Pacific, however, during certain times of the year in the absence of aeolian dust, the ocean may be the principal source of all the trace metals present in aerosol particles greater than $\sim 3 \mu m$ in diameter.

9. Enrichment of trace metals, organic substances and microorganisms occurs in the sea-surface microlayer. Toxic substances in the microlayer could adversely affect the neuston.

10. Experiments with bacteria demonstrate enrichment in the sea-surface microlayer and in the atmosphere above it, suggesting that pathogenic organisms can be transferred from sewage-polluted seawater to the atmosphere.

11. The high-molecular-weight chlorinated hydrocarbons are virtually all man-made. Such organochlorine pesticides as DDT and industrial chemicals as the PCBs have been largely banned in developed countries, because of their undesirable ecological and human-health impacts. Their use tends to be shifting to developing countries in tropical and southern hemisphere areas. As a result, their global atmospheric burden has probably not changed much in the last two decades. The net fluxes from the atmosphere to the oceans during their peak use in the northern hemisphere have been estimated at 2×10^8 g yr^{-1} for DDT and 2×10^9 g yr^{-1} for PCBs.

12. The low-molecular-weight halogenated hydrocarbons or halocarbons do not have known direct effects on the marine environment and ecosystem, but have indirect effects in that they cause erosion of the stratospheric ozone layer that offers protection against hazardous ultraviolet rays on earth. They are now largely banned in developed countries and replaced by less harmful gases in pressurized dispensers and refrigeration systems. Freon-11 (trichlorofluoro-

methane, CCl_3F), one of the most common halocarbons, was found to enter the oceans at the rate of $5.4 \times 10^9\,\mathrm{g\,yr^{-1}}$, which represented 0.5 to 1.0% of the atmospheric burden and about 2% of the total world production of F-11 during the early 1970s. This was comparable to the stratospheric sink (1%) for this halocarbon.

13. Because petroleum hydrocarbons are a complex mixture of many compounds, they present difficulties in analyses and in distinguishing them from biogenic hydrocarbons. The net flux of petroleum hydrocarbons from the atmosphere to the oceans is estimated at $0.3 \times 10^{12}\,\mathrm{g}$ out of a total of $3.2 \times 10^{12}\,\mathrm{g}$ of petroleum-derived hydrocarbons that enter the oceans each year. SEAREX studies have estimated that the total annual input of n-alkanes (n-C_{10} to n-C_{30}) from the atmosphere to the oceans is $0.04 \times 10^{12}\,\mathrm{g}$ to $0.4 \times 10^{12}\,\mathrm{g}$, which at a maximum would be 10–15% of the total petroleum input to the oceans.

14. Global fluxes of some gases across the air-sea interface have been calculated. Anthropogenic gases have a net flux from the atmosphere to the sea. A flux from the sea to the atmosphere exists for some low-molecular-weight sulphur and halide compounds produced in the sea.

15. The carbon dioxide concentration in the atmosphere has been increasing at the rate of about 1 ppmv (parts per million by volume) since good records were started at the Mauna Loa Observatory, Hawaii, in 1958, when the CO_2 concentration was measured at 315 ppmv. Ths increase represents about $2.5 \times 10^{15}\,\mathrm{g}$ of extra CO_2 carbon in the atmosphere per year from anthropogenic sources. Currently about $5.0 \times 10^{15}\,\mathrm{g\,C\,yr^{-1}}$ are added to the atmosphere from fossil-fuel burning and about $1 \times 10^{15}\,\mathrm{g\,C\,yr^{-1}}$ are estimated to be contributed from biotic sources, mainly deforestation. An imbalance of up to $4 \times 10^{15}\,\mathrm{g\,C}$ exists in the sources and sinks for CO_2. Either we are underestimating the amount of CO_2 that enters the oceans, or there is another sink for CO_2 that we have not yet uncovered.

16. Methane is another "greenhouse" gas in the atmosphere that increased from a concentration of about 1570 ppbv (parts per billion by volume) in January 1981 to 1610 ppbv by mid-1982 at the Mauna Loa Observatory, which represents a concentration increase of 2.5 ± 1.2 ppbv per month. Along with other radiatively active trace gases, such as ozone, chlorofluorocarbons and nitrous oxide, methane could have a significant impact on global temperature and climate. Of the $550 \times 10^{12}\,\mathrm{g}$ of CH_4 produced annually, most of which comes from wetlands ($150 \times 10^{12}\,\mathrm{g}$), only $13 \times 10^{12}\,\mathrm{g}$ stems from the oceans. The main sink for methane appears to be in its interaction with the atmospheric hydroxyl free radical OH.

17. Few reliable direct measurements of fluxes of substances across the air-sea boundary exist.

18. The SEAREX program is providing many definitive answers to air-sea exchange questions. Some of the difficulties in measuring fluxes of substances between the atmosphere and the sea have been identified.

Acknowledgments

Much of the information in this paper has been acquired through participation in deliberations of the GESAMP Working Group on Interchange of Pollutants between the Atmosphere and the Oceans and through its three reports. I particularly thank Dr. W. D. Garrett, the chairman of the Working Group, and Drs. R. A. Duce and P. S. Liss, who have served in the Working Group and have provided current information from such air-sea exchange studies as the SEAREX program.

References

1. NRCC: Acidification in the Canadian aquatic environment: Scientific criteria for assessing the effects of acidic deposition on aquatic ecosystems. National Research Council of Canada, Associate Committee on Scientific Criteria for Environmental Quality, Subcommittee on Water, NRCC No. 18475, 369 pp., Ottawa, Ont., Canada (1981)
2. Pierce, R.C., Whelpdale, D.M., Sheffer, M.G. (eds.): Proceedings of the Symposium on Monitoring and Assessment of Airborne Pollutants, with Special Emphasis on Long-Range Transport and Deposition of Acidic Materials and Workshop on Air and Precipitation Monitoring Networks. National Research Council of Canada, Associate Committee on Scientific Criteria for Environmental Quality, NRCC No. 20642, 502 pp. Ottawa, Ont., Canada (1983)
3. NAS: The Tropospheric Transport of Pollutants and Other Substances to the Oceans. National Academy of Sciences, Washington, D.C., 243 pp. (1978)
4. NAS: Changing Climate. Report of the Carbon Dioxide Assessment Committee, National Research Council Board on Atmospheric Sciences and Climate, National Academy of Sciences, Washington, D.C. (1983)
5. GESAMP: Interchange of Pollutants between the Atmosphere and the Oceans. IMCO/-FAO/Unesco/WMO/WHO/IAEA/UN/UNEP Joint Group of Experts on the Scientific Aspects of Marine Pollution (GESAMP). Reports and Studies No. 13, 55 pp. World Meteorological Organization, Geneva, Switzerland (1980)
6. GESAMP: Interchange of Pollutants between the Atmosphere and the Oceans (Second Report). IMO/FAO/Unesco/WMO/WHO/IAEA/UN/UNEP Joint Group of Experts on the Scientific Aspects of Marine Pollution (GESAMP). Reports and Studies No. 23, 55 pp. World Meteorological Organization, Geneva, Switzerland (1985)
7. GESAMP: Atmospheric Transport of Contaminants into the Mediterranean Region. IMO/FAO/Unesco/WMO/WHO/IAEA/UN/UNEP Joint Group of Experts on the Scientific Aspects of Marine Pollution (GESAMP). Reports and Studies No 26, 53 pp. World Meteorological Organization, Geneva, Switzerland (1985)
8. Bidleman, T.F., Olney, C.E.: Science *183*, 516 (1974)
9. Bidleman, T.F., Olney, C.E.: Nature *257*, 475 (1975)
10. Wangersky, P.J.: Organic Material in Sea Water. The Handbook of Environmental Chemistry, Vol. 1C, 25 (1984)
11. Windom, H.L.: River Inputs to Ocean Systems, p. 360, J.-M. Martin, J.D. Burton and D. Eisma (eds.). Paris, UNEP/Unesco (1981)
12. Duce, R.A., Wallace, G.T., Ray, B.J.: Atmospheric trace metals over the New York Bight. NOAA Technical Report ERL 361-MESA 4 (1976)
13. Cambray, R.S., Jefferies, D.F., Topping, G.: An estimate of the input of atmospheric trace elements into the North Sea and the Clyde Sea (1972–1973). AERE-R7733, 30 pp. United Kingdom Atomic Energy Authority AERE Harwell, HMSO, London (1975)
14. Arnold, M., Seghaier, A., Martin, D., Buat-Ménard, P., Chesselet, R.: Géochemie de l'aérosol marin au-desus de la Méditerranée Occidentale. Workshop on Pollution of the Mediterranean, Cannes, France, December 2–4, 1982 C.I.E.S.M., Monaco (1982)
15. ICES: Input of Pollutants to the Oslo Commission Area. Copenhagen, International Council for the Exploration of the Sea. Cooperative Research Report No. 77 (1978)

16. Elliott, W.P., Machta, L., Keeling, C.D.: J. Geophys. Res. *90* (1985)
17. Woodwell, G.H., Hobbie, J.E., Houghton, R.A., Melillo, I.M., Moore, B., Peterson, B.J., Slaver, G.R.: Science *222*, 1081 (1983)
18. Pacyna, J.M., Semb, A., Hanssen, J.E.: Tellus *36* B, 163 (1984)
19. Beauford, W.J., Barker, J., Barringer, A.R.: Science *195*, 571 (1977)
20. Cattell, F.C.F., Scott, W.D.: Science *202*, 429 (1978)
21. Buat-Ménard, P., Arnold, M.: Geophys. Res. Lett. *5*, 245 (1978)
22. Lepel, E.A., Stefansson, K.M., Zoller, W.H.: J. Geophys. Res. *83*, 6213 (1978)
23. Boutron, C.: J. Geophys. Res. *85*, 7426 (1980)
24. Eriksson, E.: Tellus *11*, 375 (1959)
25. Blanchard, D.C.: Prog. Oceanogr. *1*, 71 (1963)
26. Goldberg, E.D.: Comments in Geophys.: Earth Sci. *1*, 117 (1971)
27. Taylor, S.R.: Geochim. Cosmochim. Acta *28*, 1273 (1964)
28. Riley, J.P.: Chemical Oceanography, 2nd ed., p. 193, J.P. Riley and G. Skirrow (eds.), London, New York and San Francisco, Academic Press (1975)
29. Chester, R., Stoner, S.H.: Mar. Chem. *2*, 157 (1974)
30. Bertine, K.K., Goldberg, E.D.: Science *173*, 233 (1971)
31. Patterson, C.C., Settle, D., Schaule, B., Burnett, M.: Marine Pollutant Transfer, p. 23, H.L. Windom and R.A. Duce (eds.). Lexington, Mass., D.C. Heath & Co. (1976)
32. Weisel, C.P., Duce, R.A., Fasching, J.L., Heaton, R.W.: J. Geophys. Res. *89*, 11, 607 (1984)
33. Weisel, C.P., Duce, R.A., Fasching, J.L.: Anal. Chem. *56*, 1050 (1984)
34. Topping, G.: J. Mar. Res. *27*, 318 (1969)
35. Bender, M.L., Gagner, C.: J. Mar. Res. *34*, 327 (1976)
36. Bruland, K.W., Frank, R.P.: Trace Metals in Sea Water, p. 415, C.S. Wong, E. Boyle, K.W. Bruland, J.D. Burton and E.D. Goldberg (eds.), N.Y., Plenum Press (1983)
37. Boyle, E.A., Huested, S.S., Jones, S.P.: J. Geophys. Res. *86*, 8048 (1981)
38. Gordon, R.M., Martin, J.H., Knauer, G.A.: Nature *299*, 611 (1982)
39. Brewer, P.: Chemical Oceanography, Vol. 1, 2nd ed., p. 416, J.P. Riley and G. Skirrow (eds.), London, New York and San Francisco, Academic Press (1975)
40. Weisel, C.P.: Ph.D. Thesis, University of Rhode Island, Kingston, RI, 174 pp. (1981)
41. Bender, M.L., Klinkhammer, G.P., Spencer, D.W.: Deep-Sea Res. *24*, 799 (1977)
42. Klinkhammer, G.P., Bender, M.L.: Earth Planet. Sci. Lett. *46*, 361 (1980)
43. Schaule, B.K., Patterson, C.C.: Earth Planet. Sci. Lett. *54*, 97 (1981)
44. Collier, R.: (Abstract) ESO Trans. AGU *63*, 989 (1982)
45. Bruland, K.W.: Earth Planet. Sci. Lett. *47*, 176 (1980)
46. Ahmed, A.: Fluxes and processes of deposition of atmospheric sea salt to the sea surface. M.S. Thesis, Univ. of Rhode Island, 97 pp. (1978)
47. Petrenchuk, O.P.: J. Geophys. Res. *85*, 7439 (1980)
48. Fasching, J.L., Courant, R.A., Duce, R.A., Piotrowicz, S.R.: J. Rech. Atmos. *8*, 644 (1974)
49. Duce, R.A., Piotrowicz, S.R., Fasching, J.L.: Maritimes *20*, 11 (1976)
50. Piotrowicz, S.R., Duce, R.A., Fasching, J.L., Weisel, C.P.: Mar. Chem. *7*, 307 (1979)
51. Stern, A.C.: Air Pollution. 2nd ed. New York, Academic Press (1968)
52. Lantzy, R.L., Mackenzie, F.T.: Geochim. Cosmochim. Acta *43*, 511 (1979)
53. Nriagu, J.: Nature *279*, 409 (1979)
54. NAS: Manganese: Medical and Biological Effects of Environmental Pollutants. Division of Medical Sciences, National Research Council, National Academy of Sciences, Washington, D.C., 191 pp. (1973)
55. Athanassiadis, Y.C.: Preliminary air pollution summary on vanadium and its compounds. US Department of Health Education and Welfare. Public Health Service, Raleigh, N.C. (1969)
56. Rahn, K.: The chemical composition of the atmospheric aerosol. Tech. Rept., University of Rhode Island, Kingston RI, 265 pp. (1976)
57. Mroz, E.J., Zoller, W.H.: Science *190*, 462 (1975)
58. Vossler, T., Anderson, D.L., Avis, N.K., Phelan, I.M., Zoller, W.H.: Science *217*, 820 (1981)
59. Bowen, H.J.M.: Trace Elements in Biochemistry. New York, Academic Press (1966)
60. Bowen, H.J.M.: Environmental Chemistry of the Elements. New York, Academic Press (1979)

61. Robinson, E., Robbins, R.C.: Emission concentrations, and fate of particulate atmospheric pollutants. Am. Pet. Inst. Publ. 4076, 108 pp. Washington, D.C. (1971)
62. Duce, R.A., Arimoto, R., Ray, B.J., Unni, C.K., Harder, P.J.: J. Geophys. Res. *88*, 5321 (1983)
63. Arimoto, R., Duce, R.A., Ray, B.J., Unni, C.K.: J. Geophys. Res. *90*, 2391 (1985)
64. Slinn, S.A., Slinn, W.G.N.: Atmospheric Pollutants in Natural Waters, p. 23. S.J. Eisenreich (ed.), Ann Arbor, Mich., Ann Arbor Science (1981)
65. Williams, R.M.: Atmos. Environ. *16*, 1933 (1982)
66. Slinn, W.G.N.: Air-Sea Exchange of Gases and Particles, p. 299, P.S. Liss and W.G.N. Slinn (eds.), Dordrecht, Holland, D. Reidel Publishing Co. (1983)
67. McDonald, R.L., Unni, C.K., Duce, R.A.: J. Geophys. Res. *87*, 1246 (1982)
68. Arimoto, R., Duce, R.A., Ray, B.J.: SEAREX Newsletter *8*, 21 (1985)
69. Hicks, B.B., Weseley, M.L., Durham, J.L.: Critique of methods to measure dry deposition. Workshop Summary. US Environmental Protection Agency, Washington D.C. Report EPA 1600.9-80-050, 70 pp. (1980)
70. Scott, B.C.: Atmospheric Pollutants in Natural Waters, p. 3, S.J. Eisenreich (ed.). Ann. Arbor, Mich. Ann Arbor Science (1981)
71. Duce, R.A.: Vertical transport toward and into the Mediterranean. Background paper prepared for the WMO/UNEP Expert Consultation on the Atmospheric Transport of Pollutants into the Mediterranean Region. Athens, Greece, 21–25 January 1985 (1985)
72. Liss, P.S.: Chemical Oceanography, Vol. 1, 2nd ed., p. 193, J.P. Riley and G. Skirrow (eds.). London, New York and San Francisco, Academic Press (1975)
73. Liss, P.S.: Air-Sea Exchange of Gases and Particles, p. 241, P.S. Liss and W.G.N. Slinn (eds.). Dordrecht, Holland, D. Reidel Publishing Co. (1983)
74. Liss, P.S., Slater, P.G.: Nature *247*, 181 (1974)
75. Wu, J.: Science *212*, 324 (1981)
76. Wu, J.: J. Geophys. Res. *86*, 457 (1981)
77. Wu, J.: J. Geophys. Res. *89*, 8163 (1984)
78. Blanchard, D.C.: Air-Sea Exchange of Gases and Particles, p. 407, P.S. Liss and W.G.N. Slinn (eds.), Dordrecht, Holland, D. Reidel Publishing Co. (1983)
79. Monahan, E.C., Spillane, M.C.: Gas Transfer at Water Surfaces, p. 495, W. Brutsaert and G.H. Jirka (eds.). Dordrecht, Holland, D. Reidel Publishing Co. (1984)
80. Blanchard, D.C., Syzdek, L.D.: Appl. Environ. Microbiol. *43*, 1001 (1982)
81. Baylor, E.R., Baylor, M.B., Blanchard, D.C., Syzdek, L.D., Appel, C.: Science *198*, 575 (1977)
82. Baylor, E.R., Peters, V., Baylor, M.B.: Science *197*, 763 (1977)
83. Dondero, T.J., Jr., Rendtorff, R.C., Mallison, G.F., Weeks, R.M., Levy, J.S., Wong, E.W., Schaffner, W.: N. Engl. J. Med. *302*, 305 (1980)
84. Blanchard, D.C., Woodcock, A.H.: Tellus *9* (1957)
85. Lion, L.W.: The Surface of the Ocean. The Handbook of Environmental Chemistry, Vol. 1C, 79 (1984)
86. Garrett, W.D., Duce, R.A.: Air-Sea Interaction, p. 471, F. Dobson (ed.), New York, Plenum Publishing Corp. (1980)
87. Hardy, J.T.: Prog. Oceanogr. *11*, 307 (1982)
88. Waldichuk, M.: Pollutant Transfer and Transport in the Sea, Vol. I, p. 177, G. Kullenberg (ed.). Boca Raton, Florida, CRC Press, Inc. (1982)
89. MacIntyre, F.: The Sea. Vol. 5, Marine Chemistry, p. 296, E.D. Goldberg (ed.), New York, John Wiley & Sons (1976)
90. Wangersky, P.J.: Ann. Rev. Ecol. Syst. *7*, 161 (1976)
91. Lion, L.W., Leckie, J.O.: Ann. Rev. Earth Planet. Sci. *9*, 449 (1981)
92. Tsyban, A.V.: J. Oceanogr. Soc. Jap. *27*, 56 (1971)
93. Hardy, J.T.: Limnol. & Oceanogr. *18*, 525 (1973)
94. Hardy, J.T., Apts, C.W.: Mar. Biol. *82*, 293 (1984)
95. Hardy, J.T., Apts, C.W., Crecelius, E.A., Bloom, N.S.: Coast. & Shelf Sci. *20*, 299 (1985)
96. Hunter, K.A.: Mar. Chem. *9*, 49 (1980)
97. Pattenden, N.J., Cambray, R.S., Playford, K.: Geochim. Cosmochim. Acta *45*, 93 (1981)
98. Hardy, J.T., Apts, C.W., Crecelius, E.A., Fellingham, G.W.: Limnol. & Oceanogr. *30*, 93 (1985)
99. Owen, R.M., Meyers, P.A., Mackin, T.E.: Geophys. Res. Lett. *6*, 147 (1979)

100. Green, T., Houk, D.F.: Limnol. & Oceanogr. *24*, 966 (1979)
101. Broecker, W.S., Peng, T.-H.: Tracers in the Sea. Lamont-Doherty Geological Observatory, Columbia University, Palisades, New York, 690 pp. (1982)
102. Garrett, W.D.: Limnol. Oceanogr. *10*, 602 (1965)
103. Brewer, P., Frew, N., Cutshall, N., Wagner, J.J., Duce, R.A., Walsh, P.R., Hoffman, G.L., Dutton, J.W.R., Fitzgerald, W.F., Hunt, C.D., Virvin, D.C., Clem, R.G., Patterson, C., Settle, D., Glover, B., Presley, B.J., Trefry, J., Windom, H., Smith, R., Chow, T.J., Doe, B., Gram, C., Martin, J.: Mar. Chem. *2*, 69 (1974)
104. Bewers, J.M., Hartling, S., Duce, R., Walsh, P., Fitzgerald, W., Hunt, C.D., Windom, H., Smith, R., Wong, C.S., Berrang, P., Patterson, C., Settle, D.: Mar. Chem. *4*, 389 (1976)
105. Berman, S.S., Sturgeon, R.E., Desaulniers, J.A.H., Mykytiuk, A.P.: Mar. Pollut. Bull. *14*, 69 (1983)
106. Zafiriou, O.C., Gagosian, R.B., Peltzer, E.T., Alford, J.B.: J. Geophys. Res. *90*, 2409 (1985)
107. Chester, R.: Pollutant Transfer and Transport in the Sea. Vol. II, p. 67, G. Kullenberg (ed.), Boca Raton, Florida, CRC Press, Inc. (1982)
108. Goldberg, E.D.: Chemical Oceanography, Vol. 3, 2nd ed., p. 34, J.P. Riley and G. Skirrow (eds.). London, New York and San Francisco, Academic Press (1975)
109. Uematsu, M., Merrill, J., Duce, R., Chen, L.: SEAREX Newsletter *6* (1), 3 (1983)
110. Uematsu, M., Duce, R.A., Prospero, J.M., Chen, L., Merrill, J.T., McDonald, R.L.: J. Geophys. Res. *88*, 5343 (1983)
111. Erickson, D.J., Merrill, J.T., Duce, R.A.: SEAREX Newsletter *8*, 13 (1985)
112. Erickson, D.J., Merrill, J.T., Duce, R.A.: J. Geophys. Res. *91*, 1067 (1986)
113. Mahadevan, T.N., Sadasivan, S., Mishra, U.C.: Sci. Total Environ. *24*, 275 (1982)
114. Jickells, T.D., Knap, A.H., Church, T.M.: J. Geophys. Res. *89*, 1423 (1984)
115. Settle, D.M., Patterson, C.C.: J. Geophys. Res. *87*, 8857 (1982)
116. Hoffman, G.L., Duce, R.A., Hoffman, E.J.: J. Geophys. Res. *77*, 5322 (1972)
117. Buat-Ménard, P.: Air-Sea Exchange of Gases and Particles, p. 455, P.S. Liss and W.G.N. Slinn (eds.), Dordrecht, Holland, D. Reidel Publishing Co. (1983)
118. Duce, R.A., Hoffman, G.L., Ray, B.J., Fletcher, I.S., Wallace, G.T., Fasching, J.L., Piotrowicz, S.R., Walsh, P.R., Hoffman, E.J., Miller, J.M., Heffter, J.L.: Marine Pollutant Transfer, p. 77, H.L. Windom and R.A. Duce (eds.), Lexington, Mass., D.C. Heath & Co. (1976)
119. Rodhe, H., Soderlund, R., Ekstedt, J.: Ambio *9*, 168 (1980)
120. van Aalst, R.M., de Ardenne, R.A.M., Kreuk, J.F., Lems, T.: Pollution of the North Sea from the atmosphere. TNO, The Hague (1982)
121. Chester, R., Saydam, A.C., Sharples, E.J.: Mar. Pollut. Bull. *12*, 426 (1981)
122. Chester, R., Sharples, E.J., Sanders, G.S., Saydam, A.C.: Atmos. Environ. *18*, 929 (1984)
123. Seghaier, A.: Thèse de 3ème cycle, Université Paris 7, 148 pp. (1984)
124. Settle, D.M., Patterson, C.C., Turekian, K.K., Cochran, J.K.: J. Geophys. Res. *87*, 1239 (1982)
125. Buat-Ménard, P., Chesselet, R.: Earth Planet. Sci. Lett. *42*, 399 (1979)
126. Duce, R.A., Unni, C.K., Ray, B.J., Arimoto, R.: EOS *63*, 987 (1982)
127. Addison, R.F.: Can. Chem. News *38*, 15 (1986)
128. Risebrough, R., Huggett, R.J., Griffin, J.J., Goldberg, E.D.: Science *159*, 1233 (1968)
129. Cox, J.L.: US NOAA Fish. Bull. *69*, 433 (1971)
130. Young, D.R.: A synoptic survey of chlorinated hydrocarbon inputs to the Southern California Bight. Prog. Rept. to EPA. Grant. No. R801153, January 31, 1975
131. Scure, E., McClure, V.: Mar. Chem. *3*, 337 (1975)
132. Risebrough, R.W., de Lappe, B.W., Walker II, W.: Marine Pollutant Transfer, p. 261, H.L. Windom and R.A. Duce (eds.), Lexington, Mass., D.C. Heath & Co, (1976)
133. Harvey, G.R., Steinhauer, W.G., Miklas, H.: Environmental Biogeochemistry. J.O. Nriagu (ed.), Ann Arbor, Mich., Ann Arbor Science (1974)
134. Williams, P.M., Robertson, K.: US NOAA Fish. Bull. *73*, 445 (1975)
135. Bidleman, T.F., Rice, C.P., Olney, C.E.: Marine Pollutant Transfer, p. 323, H.L. Windom and R.A. Duce (eds.), Lexington, Mass., D.C. Heath & Co. (1976)
136. Hasse, L., Liss, P.S.: Gas Exchange Across the Air-Sea Interface. Report to the Second Session of the GESAMP Working Group on Interchange of Pollutants between the Atmosphere and the Oceans, West Vancouver, B.C., Canada, September 1978

137. Olney, C.E., Wade, T.L.: Chlorinated hydrocarbons in the marine atmosphere and sea surface microlayer. IDOE Workshop on Baseline Measurement, Brookhaven. N.Y., 22–26 May 1972 (1972)
138. Harvey, G.R., Steinhauer, W.G., Teal, J.M.: Science *180*, 643 (1973)
139. Villeneuve, J.P.: VII J. d'Etud. Poll. (in press)
140. Diederen, H.S.M.A., den Hartog, J.C., Hollander, Th., Kasyk, J., Schultiong, F.L.: Levels of air pollutants as measured in the period January 1969-March 1981; FLAT-project (in Dutch). Rapport CMP 81/02, TNO, Delft, Netherlands (1981)
141. Tanabe, S., Kawano, M., Tatsukawa, R.: Trans. Tokyo Univ. Fisheries No. 5, 97 (1982)
142. Giam, C.S., Atlas, E.: SEAREX Newsletter *5* (No. 2), 17 (1982)
143. Ho, R., Marty, J.C., Saliot, A.: VI J. d'Etude Poll. *39* (1982)
144. Duce, R.A., Gagosian, R.B.: J. Geophys. Res. *87*, 7192 (1982)
145. Giam, C.S., Atlas, E.: SEAREX Newsletter *5* (No. 2), 18 (1982)
146. Eichmann, R., Neuling, P., Ketseridis, G., Hahn, J., Jaenicke, R., Junge, C.: Atmos. Environ. *13*, 587 (1979)
147. Marty, J.C., Saliot, A.: Nature *298*, 144 (1982)
148. Marty, J.C.: Thèse Doctorat d'Etat. Univ. P. et M. Curie, Paris, 289 pp. (1981)
149. Lovelock, J.E., Maggs, R.J., Wade, R.J.: Nature *241*, 144 (1973)
150. CEQ: Fluorocarbons and the Environment. Report of the Federal Task Force on Inadvertent Modification of the Stratosphere (IMOS), Council on Environmental Quality, Federal Council for Science and Technology, Washington, D.C., 109 pp. (1975)
151. NAS: Halocarbons: Effects on Stratospheric Ozone. National Academy of Sciences, Washington, D.C., 352 pp. (1976)
152. Broecker, W.S., Peng, T.-H.: Tellus *26*, 21 (1974)
153. Lovelock, J.E.: Nature *256*, 193 (1975)
154. NAS: Petroleum in the Marine Environment. National Academy of Sciences, Washington, D.C., 107 pp. (1975)
155. McIntyre, A.D., Whittle K.J. (eds.): Rapp. P.–v. Réun. Cons. int. Explor. Mer *171*, 1 (1977)
156. Johnston, R.: Marine Ecology, Vol. 5, Part 3, p. 1433, O. Kinne (ed.), Chichester, U.K., John Wiley & Sons, Ltd. (1984)
157. NAS: Oil in the Sea: Inputs, Fates, and Effects. National Research Council, National Academy of Sciences, National Academy Press, Washington, D.C., 601 pp. (1985)
158. Duce, R.A., Hoffman, E.J., Fasching, J.L., Moyers, J.L.: The collection and analysis of trace elements in atmospheric particulate matter over the North Atlantic Ocean. World Meteorological Organization Publication 368, p. 370 (1974)
159. Duce, R.A., Quinn, G., Wade, L.: Mar. Pollut. Bull. *5*, 59 (1974)
160. Garrett, W.D., Smagin, V.M.: Determination of the Atmospheric Contribution of Petroleum Hydrocarbons to the Oceans. World Meteorological Organization Report No. 440, Geneva, Switzerland, 27 p. (1976)
161. Duce, R.A.: Atmospheric hydrocarbons and their relation to marine pollution. Background Papers for a Workshop on Inputs, Fates, and Effects of Petroleum in the Marine Environment, Vol. 1, p. 416, National Academy of Sciences, Washington, D.C. (1973)
162. Gagosian, R.B., Peltzer, E.T., Zafiriou, O.C.: Nature *29*, 312 (1981)
163. Gagosian, R.B., Zafiriou, O.C., Peltzer, E.T., Alford, J.B.: J. Biophys. Res. *87*, 133 (1982)
164. Schmidt, U., Kulessa, G.: The oceans as a source of atmospheric hydrogen (H_2). Abstracts ROAC Symposium, IAMAP Meeting, Hamburg, August 1981
165. Zafiriou, O.C., Alford, J., Herrerra, M., Peltzer, E., Thompson, A.M., Gagosian, R.B.: Geophys. Res. Lett. *7*, 341 (1980)
166. Ehhalt, D.H., Schmidt, U.: Pure Appl. Geophys. *116*, 452 (1978)
167. Rudolph, J., Ehhalt, D.H.: J. Geophys. Res. *86*, 11, 959 (1981)
168. Seiler, W.: Tellus *26*, 116 (1974)
169. Broecker, W.S., Takahashi, T., Simpson, H.J., Peng, T.-H.: Science *206*, 409 (1979)
170. Cohen, Y., Gordon, L.I.: J. Geophys. Res. *84*, 347 (1979)
171. Seiler, W.: The ocean as a source and sink for atmospheric trace gases. Abstracts ROAC Symposium, IAMAP Meeting, Hamburg, August 1981
172. Zafiriou, O.C., McFarland, M., Bromund, R.H.: Science *207*, 637 (1980)

173. Helas, G., Broll, A., Warneck, P.: NO_2 measurement in marine air. Abstracts ROAC Symposium, IAMAP Meeting, Hamburg, August 1981.
174. Georgii, H.W., Gravenhorst, G.: Pure Appl. Geophys. *155*, 503 (1977)
175. Granat, L., Rodhe, H., Halberg, R.O.: Ecol. Bull. (Stockholm) *22*, 89 (1976)
176. Eriksson, E.: J. Geophys. Res. *68*, 4001 (1963)
177. Graedel, T.E.: Geophys. Res. Lett. *6*, 329 (1979)
178. Barnard, W.R., Andreae, M.O., Watkins, W.E., Bingemer, H., Georgii, H.W.: J. Geophys. Res. *87*, 8787 (1982)
179. Nguyen, B.C., Gaudry, A., Bonsang, B., Lambert, G.: Nature *275*, 637 (1978)
180. Rasmussen, R.A., Khalil, M.A.K., Hoyt, S.D.: Chemosphere *11*, 869 (1982)
181. Singh, H.B., Solas, L.J., Shigeishi, H., Scribner, E.: Science *203*, 899 (1979)
182. Watson, A.J., Lovelock, J.E., Stedman, D.H.: Proc. NATO Advanced Science Institute on Atmospheric Ozone, p. 365, A.C. Aikin (ed.), US Federal Aviation Administration (1980)
183. Hunter-Smith, R.J., Balls, P.W., Liss, P.S.: Tellus *35* B, 170 (1983)
184. Gammon, R.H., Cline, J., Wisegarver, D.: J. Geophys. Res. *87*, 9441 (1982)
185. Rasmussen, R.A., Khalil, M.A.K., Gunawardena, R., Hoyt, S.D.: J. Geophys. Res. *87*, 3086 (1982)
186. Khalil, M.A.K., Rasmussen, R.A.: J. Geophys. Res. *88*, 5131 (1983)
187. Waldichuk, M.: Gas Transfer at Water Surfaces, p. 547, W. Brutsaert and G.H. Jirka (eds.), Dordrecht, Holland, D. Reidel Publishing Co. (1984)
188. Gade, H.G.: J. Phys. Oceanogr. *3*, 213 (1973)
189. Liss, P.S., Martinelli, F.N.: Thalassia Jugoslavica *14*, 215 (1978)
190. Baes, C.F., Jr., Godler, H.E., Olson, J.S., Rotty, R.M.: American Scientist *65*, 310 (1977)
191. NAS: Energy and Climate. Studies in Geophysics. National Academy of Sciences, Washington, D.C., 158 pp. (1977)
192. Craig, H., Chou, C.C.: Geophys. Res. Lett. *9*, 477 (1982)
193. Khalil, M.A.K., Rasmussen, R.A.: Chemosphere *11*, 877 (1982)
194. Barney, G.O.: The Global 2000 Report to the President. Vol. 1 and 2. Council on Environmental Quality and Department of State, Washington, D.C. (1980)
195. Zimmerman, P.R., Greenberg, J.P., Wandiga, S.O., Crutzen, P.J.: Science *218*, 563 (1982)
196. Rasmussen, R.A., Khalil, M.A.K.: Nature *301*, 700 (1983)
197. Sheppard, J.C., Westberg, H., Hopper, J.F., Ganesan, K., Zimmerman, P.: J. Geophys. Res. *87*, 1305 (1982)
198. Blake, D.R., Mayer, E.W., Tyler, S.C., Makide, Y., Montague, D.C., Rowland, F.S.: Geophys. Res. Lett. *9*, 477 (1982)
199. Ramanathan, V., Cicerone, R.J., Singh, H.B., Kiehl, J.T.: J. Geophys. Res. *90*, 55 (1985)
200. Davis, D.D., Heaps, W., McGee, T.M.: Geophys. Res. Lett. *3*, 331 (1976)
201. Crutzen, P.J., Fishman, N.: Geophys. Res. Lett. *4*, 321 (1977)

Root-Soil Interactions

P. B. Tinker

Natural Environment Research Council, Terrestrial and Freshwater Sciences Directorate
Polaris House, North Star Avenue, Swindon SN2 1EU, UK

P. B. Barraclough

Rothamsted Experimental Station
Harpenden, Herts, AL5 2JQ, UK

Summary

Root-soil interactions are physical, chemical and biological; the last two are stressed here. Changes in soil around the root follow the development of the root itself. Root density in different parts of soil is partly due to the genetic composition of the plant, partly to soil properties, including nutrient concentrations and water logging. The nutrient needs of plants, the ways in which nutrients are made available to plant roots and the mechanisms of transport are discussed, together with the value of modelling of root-soil interactions.

A large number of the most important chemical reactions in soil are mediated through the microbiological population. This includes all processes of organic material breakdown, and nearly all the important nitrogen transformations in soil. Microbes are strongly concentrated on the surface of roots and around them, largely due to the deposition of organic matter from the root into its immediate surroundings. Decomposition rates are therefore very high, but it is not clear whether nitrogen transformations are altered to an important extent. This activity may increase the soil availability of phosphate and trace elements. Symbiotic microbial systems mediate N_2 fixation and increase uptake of phosphorus and trace elements from soil.

It is important that the root also changes the pH at its surface, mainly because of differences in the cation and anion uptake rates, and this has large effects on nutrient availability. The consequences of this are not yet fully understood. Some soils have potentially toxic concentrations of heavy metals, and a decrease in pH would intensify this. A major effect of acidity is the production of aluminium ions. Different plant species have very different susceptibility to toxic metals. A number of organic toxins may also be found in soil, some produced microbially in water-logged conditions, but a few exuded from roots.

Introduction

Root-soil interactions cover such a wide range of chemical, physical and biological processes that they cannot be dealt with in any single brief review. More information can be obtained from Scott Russell (1977), Nye and Tinker (1977), Carson (1974), Tinker (1980), and Barber (1984). In this review we concentrate upon the chemical reactions involved, and especially those with implications for the environment in the sense that they determine where particular species of plants will grow successfully, and which factors or processes will affect this.

The rhizosphere is the name given to the thin zone of soil lying immediately around the root that is influenced in chemical, physical or biological ways by the presence of the root. The concept of the rhizosphere is stressed here, because the plant root system has no contact with any other part of the soil, except the root tips. The chemical and physical background of the rhizosphere is of course that of the bulk soil, on which the rhizosphere effects are superimposed. Often these bulk soil effects are dominant, and the rhizosphere complication can be ignored for practical purposes; in other cases the latter are vital. The processes range from simple fluxes of material passing to or from the root, to extremely involved biochemical reactions between soil microbes and the root surface in which there is a high degree of specific recognition. It is not possible to deal in detail with this complexity, but it is essential to include the microbiological component because so many of the most important reactions only proceed as part of the microbial metabolism in the soil.

The inner boundary of the rhizosphere may appear to be sharply defined by the root surface, but this is not always so. The surface tissues, occasionally the whole cortex, may die, and this dead material, as well as the inter-cellular spaces, may be colonized by soil microbes. It is important to remember that the rhizosphere, as defined here, is not some constant and static part of the soil. As roots grow, develop and die at different rates, individual small soil zones are continually passing through the sequence of rhizosphere development.

Root Development

The internal anatomy of the root is of little interest for this subject, and will not be discussed. The surface layer is the epidermis, or the outer cortical cell layer. Each piece of root passes through the sequence of: root tip, extension zone, root hair zone, suberization, loss of cortical cells and secondary growth. There is much variation between species – especially between monocotyledenous and dicotyledonous species – and by no means all roots will go through the whole of this sequence. Shoot/root weight ratios can range from 1:1 to around 20:1 in agriculturally important plants, but total root length is more important than weight. A root length density of 30 km m^{-2} of soil surface area has been found for winter wheat (Barraclough, 1968a). Total root weight will almost certainly be largest for forest trees, but this includes much woody thickened main root or tap root; the length of active tree roots per unit amount of soil is commonly small (Nye and Tinker, 1977). The duration of the active stage of a root is highly variable between species and conditions, and will in any case depend upon the root process taken as a criterion of activity, eg magnesium uptake ceases long before potassium uptake (Clarkson, Sanderson, and Russell, 1975).

In general, the greatest rooting densities are to be found in the topsoil. There is often an exponential distribution of root density with depth (Gerwitz and Page, 1973), but local soil conditions can strongly affect this. Some of these soil factors interact with root-soil processes. Thus poor drainage and waterlogging of soil invariably result in anaerobic soil, since the normal metabolism of soil microorganisms depletes oxygen in the soil air and there are no air-filled pore spaces through which oxygen can enter. Root metabolism alters, and a new set of processes starts in the soil which eventually produces toxins such as Fe^{++} or H_2S. All these processes greatly restrict root extension. The density of roots in waterlogged soil is normally much less than in aerated parts of the same profile, hence the many special chemical reactions in these conditions are proportionally less important to the majority of plant species, and will not be further discussed here. Roots of some species are less affected, but only because the roots themselves act as channels for oxygen supply to the rhizosphere (Armstrong, 1978).

In nutrient deficient soils, root (and shoot) growth is reduced in absolute terms but the root/shoot ratio is normally increased. The exponential distribution of rooting density is however maintained (Barraclough, unpublished). Arable soils may not be nutrient-deficient in total, but the nutrients are usually distributed non-uniformly throughout the rooting zone. In this case localised stimulation of root growth and activity can occur. In solution culture, Drew (1975) and Drew and Sakar (1978) found that the extension of axes was little influenced by the immediate concentrations of N, P and K. However, the growth and absorbing and transporting powers of lateral roots were enhanced in zones where concentrations of nitrate and phosphate (but not potassium) were locally high, if the remainder of the roots were deficient. This localised stimulation was able to compensate for the deficient nutrient supply to the rest of the system so that growth was not affected, though clearly there must be limits this response.

Soil pores wider than about 50 μm are drained of water at field capacity, and root diameters are usually much greater than this (typically 200–400 μm for cereals). Most root growth will therefore be in air-filled pores and root-soil contact has to be maintained through thin water films held between the surfaces of soil particles and root cells. Tinker (1979) estimated the effect of partial root-soil contact on the uptake of water (and thus on mass flow transfer) and De Willigen (1984) more fully modelled the effects of root-soil contact on uptake of nutrients and water. In such circumstances root hairs may be of great importance for soil-root contact.

Root hairs are extensions of epidermal cells with lengths in the range 0.1–5 mm and diameters 0.005–0.025 mm, and they may consequently greatly modify the root surface geometry. They may only survive for a few days, but the duration, length and density of root hairs per unit root length vary both within and between species, and are also very sensitive to external conditions. Foehse and Jungk (1983) found that low concentrations of nitrate and phosphate in solution culture produced longer and more root hairs in rape, spinach and tomato, and that their formation depended on the P content of the shoot rather than that of the root. However, in the work of Ewens and Leigh (1985), the length of root hairs of wheat was affected mainly by pH and the concentrations of calcium and nitrate, whereas phosphate concentrations had little or no effect. Mackay and Barber (1985) reported that root hair growth of corn in soil was affected more by soil moisture than by soil phosphate. It appears that much more work is needed before the mechanisms which determine root hair behaviour are understood.

The importance of root hairs for phosphate uptake in any given situation must depend on their length in relation to the spread of the phosphate depletion zone around a root. There are cases in which this appears to be important in practice, for example, white clover genotypes with long root hairs were considerably more efficient in phosphorus uptake from soils than genotypes with short hairs (Caradus, 1982).

Root hairs can also be important for potassium uptake. For a three-fold increase in the volume of the root hair cylinder of different species, potassium inflow increased five-fold (Claassen and Jungk, 1984).

Chemical Interactions Between Roots and Soil

Nutrient Requirement of Plants

The most obvious way in which roots interact with soil is in the uptake of mineral nutrients. The total mineral content of green plant material is of the order of a few percent, the following elements being recognized as essential for the growth of higher plants: N, P, K, S, Ca, Mg, Fe, Mn, B, Zn, Cu, Mo and Cl. The relative amounts of essential nutrients required by plants are fairly constant; in non-deficient plants the N and K contents of green plant material are normally about 10 times those of P, S, Ca and Mg which in turn are about 10–1000 times those of the remaining essential elements (Table 1). Despite this, chloride may be present in the plant tissue at concentrations of several percent if it is freely available in the soil.

Table 1. Nutrient concentrations in the dry matter of healthy plants (Mengel and Kirkby, 1982)

Nutrient	Concentration in plant dry matter
N	2–5%
K	2–5%
P	0.3–0.8%
S	0.1–1%
Ca	0.3–3%
Mg	0.1–0.5%
Fe	100 ppm
Mn	100 ppm
B	30–200 ppm
Zn	20–100 ppm
Cu	2–20 ppm
Mo	1 ppm

Plants obtain most of their mineral nutrients from the soil solution. Non-metals are taken up as the anions NO_3^-, $H_2PO_4^-$, HPO_4^{2-}, SO_4^{2-}, Cl^-, except that B is taken up as H_3BO_3 and N may also be taken up as NH_4^+. The metals are taken up as the simple cations: K^+, Ca^{2+}, Mg^{2+}, Fe^{2+}, Mn^{2+}, Zn^{2+}, Cu^{2+}, except for molybdenum. Uptake is selective, in that the absolute concentrations and ratios of nutrients in xylem sap or in the vacuole differ widely from those in the soil solution, and that those in the plant tend be more constant. However, non-essential and even toxic elements are also absorbed. Uptake may be "active" or "passive", in the sense that in active uptake, ions are transported against an electrochemical potential gradient between the soil solution and the interior of root cells, with expenditure of metabolic energy by the plant.

Many native species have been naturally selected primarily for survival in deficient environments, and their strategy is to grow and absorb nutrients slowly (Rorison, 1969; Chapin, 1982). In this case nutrient cycling between plant and soil is especially important for maintaining the supply of nutrients. Crop plants, on the other hand, have been selected for rapid growth, and hence have large nutrient demand; nutrients are also lost from the system when crops are harvested, so that fertilizer is necessary for continuing high crop yields. A large wheat crop, for example, with 20 t dry matter ha^{-1} will remove the following amounts (kg ha^{-1}) of nutrients from the soil: N, 280; P, 50; K, 300; Ca, 50; S, 50; Mg, 30; Fe, 5; Mn, 0.5; B, 0.4; Zn, 0.4; Cu, 0.2; Mo, 0.02 (FAO, 1984). If straw is returned to the soil, the loss (kg ha^{-1}) in 10 t ha^{-1} grain would be: N, 200; P, 35; K, 50; Ca, 5; S, 15; Mg, 10; Fe, 0.4; Mn, 0.3; Zn, 0.3; Cu, 0.05 (McGrath, 1985). When the crop is growing rapidly, daily uptake rates of 5, 0.5 and 5 kg ha^{-1} d^{-1} are likely for N, P, and K respectively (Barraclough, 1986a).

Amounts, Distribution and Mobility of Nutrients in Soil

The total content of a nutrient element in the soil is determined by the soil parent material, climate, vegetation and fertilizer practices. The chemical composition of the soil varies with depth, and the surface horizon is generally the most fertile because it is here that plant remains are broken down and fertilizer is applied. The total amount of a nutrient element present in the soil is rarely available for immmediate uptake by plant roots. Nutrients are absorbed from the soil solution, but the amounts present there (Table 2) are inadequate to sustain growth for long. It is therefore essential that the soil solution is replenished with nutrient ions by dissolution, desorption and exchange from the soil solid phase. Three categories of solid are of interest: completely soluble salts (eg nitrates, chlorides), sparingly soluble salts and complexes ($CaSO_4$, $CaCO_3$, $CaHPO_4$, $Ca_3(PO_4)_2$, $MgCO_3$, $FePO_4$, $AlPO_4$, Fe- and Cu-chelates) and exchange complexes (clays and organic matter with inherent or pH-dependent negative charges balanced by exchangeable cations). There is continuing two-way exchange between solid and liquid phases, and if any chemical species is withdrawn from the solution, equilibrium will be re-established, at widely varying rates, by transfer from soil solids.

Nutrient movement in soil is largely mediated in the soil water. Nutrients can move by mass flow of water (drainage or flow induced by plant uptake or soil drying) and by diffusion in the soil water. Nitrate and chloride ions normally interact negligibly with the soil solids, hence they are very mobile, and in humid climates these ions are readily leached into subsoils and ultimately into aquifers and water courses. The rate of leaching depends on many environmental factors, and simulation models of varying degrees of complexity have been developed (Addiscott and Wagenet, 1985). The depth of rooting determines whether nitrate leached

Table 2. Typical total plant nutrient contents in the top 20 cm of soil and their concentrations in the soil solution

Nutrient	Soil (t/ha)	Soil solution (M)
N	12	10^{-2}–10^{-3}
P	15	10^{-4}–10^{-6}
K	120	10^{-3}–10^{-4}
S	1	10^{-3}–10^{-4}
Ca	300	10^{-2}–10^{-3}
Mg	15	10^{-3}–10^{-4}
Fe	60	10^{-7}–10^{-8}
Mn	9	10^{-4}–10^{-6}
Zn	0.9	10^{-6}–10^{-8}
Cu	0.2	10^{-6}–10^{-8}
Mo	0.1	10^{-7}–10^{-8}
B	0.6	10^{-4}–10^{-6}

Sources: Mengel and Kirkby, 1982; Barber, 1984

Table 3. Measured diffusion coefficients of ions in soil ($cm^2 s^{-1}$)

NO_3^-	0.5–5.0×10^{-6}
Cl^-	0.3–13.5×10^{-6}
Na^+	0.5–2.2×10^{-6}
NH_4^+	0.4–3.0×10^{-7}
K^+	1.0–24.0×10^{-7}
Mg^{2+}	0.3–64.5×10^{-7}
Ca^{2+}	0.3–3.3×10^{-7}
Zn^{2+}	2×10^{-7}–3.1×10^{-10}
Mn^{2+}	3.3–22.0×10^{-8}
MoO_4^{2-}	4.6–51.0×10^{-8}
$H_2PO_4^-$	4.0×10^{-8}–2×10^{-11}

Source: Barber, S. A., 1974

Table 4. Root-mean-square displacement (x) of diffusing ions in one day

D ($cm^2 s^{-1}$)	X (mm)
10^{-5}	13.1
10^{-6}	4.2
10^{-7}	1.3
10^{-8}	0.42
10^{-9}	0.13
10^{-10}	0.042
10^{-11}	0.013

into the subsoil is still available for uptake. All the nitrate to a depth of 90 cm can be taken into account when planning fertilization rates for cereals (Wehrmann et al., 1982), but effective rooting depths as small as 15 cm have been measured for other crops such as lettuce (Burns, 1980). Little is known about rooting depths of natural vegetation. Leached nitrate or other anions must be accompanied by cations, mainly Ca^{2+}, and this causes soils gradually to become acidic. However, leaching prevent a build up of salts in the surface horizon. In arid or semi-arid regions high rates of evaporation prevents through leaching, and induce transport of salts from the subsoil to the surface soil, where they eventually give rise to a salinity problem (Poljakoff-Mayber and Gale, 1975).

At the other extreme of ionic mobility, phosphate and the trace elements interact strongly with the solid phase and are present in the soil solution at very low levels, of around 10^{-5} M, so that their net rate of movement is extremely slow. Cooke and Williams (1970) found that fertilizer P applied at a rate of 33 kg ha^{-1} yr^{-1} for 100 years had only penetrated 20 cm below the plough layer on a heavy arable soil. Diffusion is the main mechanism for transporting P in soil (Nye and Tinker, 1977; Nye, 1979). Diffusive flux depends on the diffusion coefficient and the concentration gradient. In free solution, diffusion coefficients for all simple ions are in the range 1–2×10^{-5} $cm^2 s^{-1}$. In the soil, however, effective diffusion coefficients may range over five orders of magnitude (Table 3), because of the smaller cross-sectional area of the soil medium available for diffusion, the tortuous pathway that ions must follow in soil, and the sorption of the ion on the solid phase. Simple theory makes no allowance for diffusion in the exchange phase, though evidence has been found for this with Na^+ in structureless clays and in a naturally aggregated soil (Staunton and Nye, 1983).

The significance of the diffusion process for ion movement in soil can be simply gauged by calculating the root-mean-square displacement x of an individual ion undergoing diffusion; $x = \sqrt{2Dt}$, where t is time and D is the diffusion coefficient (see Table 4).

Transport Mechanisms in Relation to Nutrient Supply to Roots

Dissolved nutrients must be transported to root surfaces by mass flow and diffusion prior to uptake (Barber, 1974; Nye and Tinker, 1977). Water moves to roots as a result of transpiration and all solutes are carried with it. If plant demand for nutrients exceeds their rate of arrival at root surfaces by mass flow, the concentration at the root surface falls relative to that in the bulk soil solution, and ions then diffuse down the resulting concentration gradient. The size of the gradient, and hence the diffusion flux, depends on the balance between root uptake rate, rate of transport by mass flow and diffusion, and rate of release from the solid phase. Diffusion is likely to be important at all times when transpiration rates or solute concentrations are low, though simple calculations show that all of the Ca^{2+}, Mg^{2+}, and S needed by a crop over a whole season, a large part of the N but very little of the P and K, could *in theory* be supplied by mass flow (Barber, 1984). If the transport by mass flow exceeds the uptake rate, the concentration at the root surface rises (Riley and Barber, 1970; Nye and Tinker, 1977). Root geometry (length, branching pattern and density, root hairs) determines the volume of soil in close proximity to roots and it is an important factor in determining the maximum rate at which immobile nutrients like P can be taken up. The rate of diffusion of ions in the rhizosphere is not necessarily the same as that in the bulk soil, since the root modifies the soil in this zone (see appropriate Section).

From actual field measurements, 75% of the total N uptake by a spring wheat crop (Strebel et al., 1980) and a winter wheat crop (Barraclough, 1986 b) growing under humid conditions was supplied by diffusion. However, the relative contribution of mass flow and diffusion in supplying nutrients changes during the growing season in response to the weather and fertilizer applications (Gregory et al., 1979). So far no comparable measurements have been made under natural vegetation. It is likely that soil solution concentrations are much less than in agricultural soils, hence the mass flow process is likely to be less important, and diffusion may be the dominant process for most ions. At low soil moisture contents, the rate of supply of nutrients by both mass flow and diffusion is substantially reduced, leading to reductions in plant nutrient contents (Sharpley and Reed, 1982). Under such conditions nutrient deficiency as well as water stress limit plant growth, as has been shown in work with rain shelters in the field and also in relation to the effects of mycorrhiza.

Differences in the concentrations of ions at the root surface and in the bulk soil solution during nutrient uptake have been confirmed by autoradiography, by separation of rhizosphere soil from roots, and by thin-sectioning of soil held against roots. With these methods it has been demonstrated that plant roots can decrease the concentration of phosphorus in the soil solution at the root surface to below 1 µM (Hendricks et al., 1981), and of potassium to 2 µM (Jungk et al., 1982). Steep gradients of potassium concentration were found up to a distance of about 3 mm from young maize and rape roots. In addition, the depletion of potassium within the rhizosphere caused enhanced release of potassium from non-exchangeable sites on the clay fraction of the soil up to twice that of the original amount of exchangeable K.

Experiments on nutrient uptake from solution invariably show that the flux into roots is concentration dependent. The most recent solution culture studies using rapid flow rates and very dilute concentrations have confirmed this, but have also revealed that maximum plant growth rates can be achieved at very low solution concentrations, of the order of 1 μM, provided that these can be maintained (Clement et al., 1978; Wild and Breeze, 1981; Breeze and Wild, 1984). These studies and others (Glass and Siddiqi, 1984) have shown clearly that the uptake properties of plants are not constant, but adjust themselves to the supply of nutrients and the state of the plant.

Mathematical Modelling of Nutrient Uptake from Soil

Mathematical modelling has been employed extensively in work on nutrient uptake. The technique is important because of the complex interplay of soil, plant and other environmental variables, so that only modelling methods can allow an understanding of the system as a whole. Conversely, the success of modelling techniques gives some confidence that our understanding of nutrient uptake processes is essentially correct.

Several simple mathematical transport models have been developed which assume the radial transport of ions by mass flow and diffusion through the soil to a smooth, cylindrical sink (the root), and these allow changes in the concentration gradient around the root with time to be calculated. Good agreement between predicted and observed concentration profiles around roots in simplified situations has been found, showing that the underlying theory is essentially correct (see Nye and Tinker, 1977). This basic model has subsequently been elaborated for the prediction of nutrient uptake by whole root systems growing in soil (Brewster and Tinker, 1970; Baldwin et al., 1973; Claassen and Barber, 1976; Barber and Cushman, 1981). These models have been verified for N, P and K uptake over short periods by plants growing in pots of soil (see Nye and Tinker, 1977; Barber, 1984), the soil and plant parameters required in the models being independently measured. For satisfactory prediction of uptake over any extended period of time, it is necessary to take root competition into account which is normally done by assigning a limited cylinder of soil to each root. As a rough guide, if the mean inter-root distance is greater than $\sqrt{Dt}$, a root system can be regarded as composed of isolated single roots.

The models have also been used to conduct sensitivity analyses of nutrient uptake to different soil and plant parameters. Root extension rate and root radius were the most sensitive parameters influencing phosphorus uptake. Phosphorus uptake was also more sensitive to the soil supply parameters (soil solution concentration, diffusion coefficient, buffer power) than to root physiological parameters (maximal influx, Michaelis-Menten constant) (Silberbush and Barber, 1983).

The modelling of phosphorus uptake has presented most problems, because predicted uptakes have not always agreed with observed uptakes. Phosphate has very low mobility and is absorbed from a very localised region. Modifications of the rhizosphere soil and root morphology appear to be important, and models based on roots as inert, smooth cylindrical sinks are not satisfactory. The effective morphology of roots may be altered by the presence of root hairs and fungal hyphae, and the equilibrium concentration of ions in the soil solution close to roots may be altered by root induced pH changes and the release of exudates.

It is therefore not surprising that Itoh and Barber (1983) found that phosphorus uptake by root hairs had to be included in their model for good agreement between observed and predicted P uptake for tomato and Russian thistle, both of which had long root hairs. The shorter hairs of other species gave no enhancement of phosphorus uptake over that calculated for a simple root.

Practical Use of Modelling

The ultimate aim of the modelling exercises is to obtain a quantitative understanding of the major processes involved in plant nutrition and to be able to model nutrient uptake by crops precisely, so that fertilizers can be used as efficiently as possible. Barraclough (1986 b) used a model to calculate the soil solution concentrations needed to sustain the observed maximal inflows of a winter wheat crop. Assuming only diffusive transport of nutrients in a moist soil, 165 μM N, 13 μM P and 55 μM K were necessary, all relatively low values compared with the usual concentrations to be found in fertile arable soils. In a dry soil however, these concentrations were increased 7-fold, and if only young roots in the plough layer were assumed to be active, the required concentrations were increased 15-fold. So far such models have hardly been tested under field conditions. Schenk and Barber (1980) found that predicted phosphorus uptake by maize was only 67% of the observed uptake, probably because no allowance was made for root hairs or mycorrhizae in the model used. Silberbush and Barber (1984), on the other hand, found good agreement between observed and predicted potassium uptake by field-grown soyabeans, but only when the initial soil solution concentration was 52 μM; at 8 μM predicted uptake was less than observed. The models appear to be generally correct, but there is still considerable scope for improvement.

The main problems in modelling the field situation are the heterogeneity of natural soil profiles, and the prediction of root growth. Differences occur over short distances in nutrient and water contents, buffer power, diffusion coefficient and root growth, so the total soil volume has to be broken down into units and uptake from each modelled separately. This presents no problems in principle for the modelling, but considerable problems for experimental verification. There is little information about the longevity of root activity in the field, or about the length of root that is actually in contact with soil at any given time, but root characteristics must be important for nutrient uptake. Jungk et al. (1982), could only account for differences in the K content of rape, maize and onion shoots by taking the uptake rate per cm of root, the root/shoot ratio and the mean age of the root system into consideration.

Microbiologically Mediated Chemical Transformations

The most common definition of the rhizosphere (Rovira, 1979) is in terms of the increased microbial population, which is commonly increased by a factor of up to 10 (see Newman and Watson, 1977). The cause lies in the additional supplies of organic substrate supplied by the root, both as soluble and diffusible compounds, and as insoluble debris. There is no fundamental difference between the microbial species in the bulk soil and the rhizosphere, except that rapidly multiplying species such as Pseudomonads are relatively more numerous. This conclusion is based upon simple identification by plating out soil extracts, which recovers only a small fraction of all soil microbial species (Macdonald, 1986).

Root Exudates

Almost every group of compounds found in plants have also been detected in root exudates (Hale et al., 1971). The most common are sugars, carboxylic acids and amino acids. Some of these soluble materials are released when cells in the cortex become senescent and damaged, some may be the product of bacterial metabolism on the root surface, but undoubtedly much of it is exuded from healthy roots – particularly the part near the apex – as a normal growth process. The main uncertainty at present is about the total quantity of material. Soluble material has rarely been detected in amounts of more than 1%–2% of total plant weight, but the use of 14C labelling of plants, with subsequent analysis of soils and measurement of below-ground CO_2 production, suggests that total amounts of C lost below ground are in the region of 20%–40% of total plant weight (Newman, 1985; Whipps and Lynch, 1985). Trolldenier (1979) has suggested that the mineral nutrition of the plant has a significant effect on exudation. The difficult problem is to determine how much of this CO_2 is produced by direct root respiration, and how much by microbial metabolism of carbon compounds deposited in the rhizosphere. The simple method of distinguishing these processes – to work in sterile soil – is not acceptable because the presence of microbes actually increases the rate of exudation.

As some bacteria inhabit the intercellular spaces within the root (Lethbridge and Davidson, 1983), the distinction between exudation and CO_2 production is difficult to make even in principle. This is also complicated by the fact that symbiotic organisms such as *Rhizobia, Frankia* and mycorrhizal fungi appreciably increase the rate of metabolism of the root when it becomes infected. At present it is therefore impossible to reach a clear conclusion about the amount of carbon lost from the root in forms other than CO_2, but it seems reasonable to assume that the maximum of 1%–2% of plant weight found as soluble exudates in solution culture may also be about correct in soils. Another 5%–10% may be lost as root cap material, decaying root hairs, and fine roots and sloughed-off cortical or epidermal cells, and the rest is respiratory CO_2. This would agree with the conclusion of Helal and Sauerbeck (1984) that about 7.5% of total fixed carbon was deposited as organic material in the rhizosphere of maize. The material is far too diverse to consider its breakdown in detail by the soil microbial population – for references to the general metabolism of organic materials by the "biomass" (including soil microfauna) see Brookes et al. (1985). Some materials such as lignin are resistant, and only recently have these processes begun to be understood. The discovery of a lignin-degrading enzyme and studies on the genetics of its production are now beginning to explain these processes. The breakdown of root material, as with other organic tissues, eventually results in the formation of the stable part of soil organic matter, or humus. The presence of this material converts a mixture of mineral grains into functional soil possessing biological activity, nutrient supply and structural stability, and as such it is of enormous environmental importance.

Occasionally there is an indication that a specific compound with identifiable effects may be released, and this could have great practical importance. If it were possible to spray onto crop leaves fungistatic compounds which were translocated and exuded into the rhizosphere, this could have important applications in disease control. So far no natural compounds with specific properties have been identified, though Helal and Sauerbeck (1984) detected increased levels of phosphatase enzymes following deficiency of phosphate.

Nitrogen Transformations

The processes in the nitrogen cycle mediated by microbes are: fixation, mineralisation, nitrification, denitrification. All of these processes occur in the bulk soil, and undoubtedly also in the rhizosphere. The question is whether the particular conditions induced by the presence of the root cause a change of rate or of end products.

Fixation is either symbiotic (see below) or associative. In the latter, bacteria of various genera (*Clostridium, Azotobacter, Azospirillum*) fix N_2 gas in their tissues. This occurs at a slow rate in most soils, but in the last decade there has been controversy over whether the fixation in the rhizosphere is accelerated to a point where it yields a large net increase in nitrogen for plant uptake (Giller and Day, 1985). The reasons for expecting high fixation in the rhizosphere are (a) low mineral nitrogen levels because of plant uptake, so that fixation is not inhibited; (b) the availability of relatively large amounts of energy in the carbon compounds lost from the root. The latter contributes towards the high energy cost of N_2 fixation, and also tends to produce locally anaerobic conditions, which are necessary for the nitrogenase enzyme to remain functional. Despite these advantages, and earlier reports of large fixation rates, the evidence for this in field circumstances is slight, and there is apparently no comparison with the N_2 fixation rates achieved by symbiotic systems. If the rates are large anywhere, it is likely to be in the tropical grasses with high growth rates. However, low rates of fixation of perhaps 10–20 kg N ha^{-1} yr^{-1} may still be of importance for nitrogen accumulation in natural vegetation.

Mineralization occurs when nitrogen-containing carbon compounds in soil are used as microbial substrates, and the surplus nitrogen they contain is execreted as ammonium compounds. The process decreases with the C/N ratio of the substrate, ceasing when this ratio rises above about 30.

Mineralization is effected by a very wide range of microorganisms – in fact by all heterotrophic organisms, as ammonia is simply their waste product. It may occur at an accelerated rate in the rhizosphere simply because the saprophytic population is increasing rapidly, but the mean C/N ratio of the materials released is not likely to be particularly low. It is more likely that ammonia is liberated during the later phases of rhizosphere development, when the total rhizosphere biomass is decreasing along with root activity (Newman and Watson, 1977).

Nitrification is effected by a small number of specialised bacteria in aerobic conditions – *Nitrosomonas* oxidises ammonia to nitrite, *Nitrobacter* oxidises nitrite to nitrate. As the bacteria are autotrophic, there is no reason to expect their numbers to increase markedly in the rhizosphere, where ammonium concentrations are likely to be low.

Denitrification occurs in anaerobic environments when the nitrate ion acts as an electron acceptor, and denitrifying organisms use it as such in their metabolism. At low redox potentials N_2 or N_2O gas is released (Stefanson, 1973). There is some evidence that the rate of this process may be increased by plant roots, but the evidence is controversial at present. The reason for expecting it to occur is the supply of organic substrates from the roots which may generate anaerobic conditions if the soil is almost waterlogged. When the air-filled porosity is less than around 10% of soil volume, the diffusion rate for oxygen becomes very small and the presence of decomposable materials will generate local denitrification. Thus it is most likely to occur when roots die because of anaerobic conditions.

Siderophores

Much of the chemistry of heavy metals in soil is dominated by their tendency to form chelates (Tinker, 1986). The presence of complex-forming compounds in the soil solution determines the solubility, diffusibility and availability of trace metals such as Cu, Zn, Mn and Fe. Chelating compounds are generated in a non-specific way by the general metabolism of the soil population, but soil and rhizosphere microorganisms can also produce specific compounds with high affinity for particular metals (siderophores) and roots can excrete compounds with similarly specific properties (Römheld and Marschner, 1986). It seems likely that siderophores are produced particularly rapidly in the active zone of the rhizosphere, both from roots and microbes.

Root Surface Enzyme Activity

There is appreciable enzymic activity in soils, the enzymes being in the main derived from dead microbes (Burns, 1978). However, root surfaces also possess active enzymes, the most investigated of which are phosphatases. It has often been shown that roots in solution culture will cause organic phosphates to decompose to orthophosphate, and the activity appears to be particularly marked when the roots carry ectomycorrhizal sheaths (Bartlett and Lewis, 1973). There is much argument over whether these phosphatases have an effective function, i.e. whether they decompose soil phosphate esters and so enhance the supply of inorganic phosphate ions to the root. However, there is normally a very low concentration of phosphate esters in the soil solution, and their practical importance is still not clear.

Symbiotic Systems

The main symbioses are with the nitrogen fixing organisms – the bacterium *Rhizobium* and actinomycetes such as *Frankia* – and the various fungi which form mycorrhizas. The biochemistry of the nitrogen fixation process is fairly well understood, but much remains to be discovered about the selectivity and interaction mechanisms. Apart from the initial infection step, there is little direct interaction between the soil and the symbiotic parts of the root (the nodules). However, soil acidity is damaging to the Rhizobium, and can prevent useful infection. The interplay between acidity effects on the host and on the Rhizobium is very complex (Munns, 1986).

The taxonomy and biology of the mycorrhizas has been extensively reviewed (see Harley and Smith, 1982). It must be stressed that the great majority of wild and cultivated plants are mycorrhizal, and that this is their normal condition. The etcomycorrhizal fungi produce sheaths or coverings on many roots, and cause changes in root morphology. As most active roots are wholly enclosed by a sheath or mantle of fungal tissue, the root proper is hardly in contact with the soil at all, and the chemical interactions with the soil must be seen as either being with the fungus, or at least mediated by it. There is however no real reason to believe that these interactions are different in principle from those with a simple root. The endomycorrhizas have no obvious fungal material on the roots, except for a network of fine hyphae extending into the soil from the root. The most important process is the uptake of ions, in particular phosphate, by the distributed hyphal system which is a particularly effective diffusion sink for immobile ions. This is achieved through the increased and better distributed absorbing surface provided by the hyphae in much the same way as root hairs, rather than to P "solubilizing" powers possessed by the fungus (Sanders and Tinker, 1973). The phosphorus uptake kinetics of external mycorrhizal hyphae are largely unknown, as are their total length, density and growth rate in any given situation (the usual measurement is of the percentage of root length infected). There are indications that the affinity of the fungal hyphae for phosphorus is higher than that of root cells (Cress et al. 1979). As yet none of the published models include nutrient uptake by mycorrhiza. In all mycorrhizae tested so far it has been found that orthophosphate is converted in large part to polyphosphates (Tinker, 1975). There is almost certainly an important storage function for polyphosphate in the large amounts of mycelium in ectomycorrhizas, and in endomycorrhizas it appears to be the main form translocated along the hyphae.

The question of whether root exudation rates are increased or decreased by mycorrhizal infection has not been properly settled by comparisons of similar plants. The proposals of Ratnayake et al. (1978) on the function of exudate as the substrate for the mycorrhizal fungi would certainly suggest that less is lost from a mycorrhizal root, but direct evidence for this is slight. The ectomycorrhizal fungi themselves produce compounds with phytohormonal effects (Slankis, 1973), and these are believed to be the causes of the changes in root morphology accompanying these infections.

Infections with pathogenic organisms, or with pests such as nematodes, can effect massive changes in root morphology and inefficiency in uptake of nutrients and water (Schippers and Gams, 1979), but it has not proved possible to correlate this loss of efficiency to the degree of infection.

Chemical Modification of the Rhizosphere by Roots

Exudation

Such processes are known to be important for plant nutrition in some situations; for example, Bhat et al. (1976) found that rape roots took up more P than could be accounted for by diffusion theory even when root hair activity had been taken into account. Since rape is not infected by mycorrhizae, this strongly suggests that P was being "solubilized" in the rhizosphere by some root-induced chemical change.

If the soluble part of a root exudate (immediately, or after metabolism by microorganisms) has the ability to complex aluminium, iron and/or calcium, the activities of these ions in the soil solution will be decreased, and any phosphate salts of these cations will tend to dissolve. Alternatively, some hydroxycarboxylic acids will displace sorbed phosphate ions from oxide surfaces. It is still very much an open question whether these processes are important. Tinker (1980) and Nye (1984) have made simple calculations on the maximum effect, and have shown it to be small on the most likely assumptions. However, Gardner et al. (1983) have found very large exudations of citric acid from lupins, which could well have caused such an effect. More measurements of carboxylic acid loss into the rhizosphere, and preferably direct evidence of its effect, are needed.

pH in the Bulk Soil and Rhizosphere

Soil pH depends on the nature of the parent material, and the intensity of weathering and leaching. It can also be manipulated by cultivation practices. A soil becomes acidic when basic cations such as Ca^{2+}, Mg^{2+} and K^+ are leached from the profile faster than they are released by the weathering of the parent minerals. If so, these cations are replaced by Al^{3+} and H^+ ions on the exchange sites of clay minerals and in the soil solution. The presence of free $CaCO_3$ in the soil delays this process and buffers pH above neutrality. The main acidifying processes in the soil are:-

(a) The microbial breakdown of organic matter releases CO_2 and SO_4^{2-} into the soil, as well as ammonia. The microbially-mediated nitrification process increases soil acidity by oxidation of ammonia. CO_2 dissolved in the soil water is an important acidifying agent in soils of initially alkaline pH, as it forms carbonic acid and then bicarbonate ions.

(b) The leaching of soluble anions in the soil solution (NO_3^-, HCO_3^-, Cl^-, SO_4^{2-}) is accompanied by an equivalent amount of cations (usually Ca^{2+}).

(c) SO_2 and NO_2 released by combustion of fuel and industrial processes oxidise to H_2SO_4 and HNO_3 in the atmosphere, on leaves and at the soil surface. Acid rain in industrial regions can have a pH of 4 or less, and is a cause for much concern (Bache, 1980).

(d) Oxidation of any iron pyrites in soils formed under earlier reducing conditions gives rise to H_2SO_4 and $Fe(OH)_3$:

$$2\,FeS_2 + 7O_2 + 2H_2O = 4SO_4^{2-} + 4H^+ + 2Fe^{2+}$$
$$2\,Fe^{2+} + \tfrac{1}{2}O_2 + 5H_2O = 2Fe(OH)_3 + 4H^+$$

(e) Some fertilizers such as ammonium sulphate are strongly acidifying. Total
    annual losses of calcium from temperate arable land are thought to be up to
    the equivalent of 1 t $CaCO_3$ ha^{-1} (Archer, 1985).

Acidification is thus a natural process in humid climates when there is much
through-drainage, but it can be accelerated by various human agencies. Acid soil
conditions generally affect plant growth indirectly through changes of nutrient
availability, microbial activity or by release of elements in toxic amounts into the
soil. Root growth is only negligibly affected by pH in the range 4–8 if sufficient
Ca is available and if excess toxic ions such as Al^{3+} are absent, but a pH of less
than 4 may cause direct impairment of root growth and functioning, when root
cell membranes may be damaged.

The pH of the rhizosphere soil may differ from that of the bulk soil by up to
1 pH unit (Nye, 1986). Several methods have been used to determine such pH
changes, including direct measurements with antimony microelectrodes
(Römheld and Marschner, 1984). Roots (and microbes) exude organic acids and
release CO_2 during respiration. However, the amount of organic acids is unlikely
to be sufficient to have a large effect. Much more CO_2 than organic acids would
be produced by roots, but the CO_2 would rapidly diffuse away from the root
under aerobic conditions, and would have no significant local effect.

The main process for altering rhizosphere pH is undoubtedly the uptake of
different amounts of anions and cations by plants, which are then obliged to release HCO_3^- or H^+ to preserve electroneutrality (Tinker, 1980). In most circumstances where plants use NO_3-N as the nitrogen source, more anions than cations
are taken up and consequently roots should release OH^- or HCO_3^- rather than
H^+, so pH in the rhizosphere should increase compared with that in the bulk soil.
This has been found in several studies, with increases up to 0.5 pH unit being observed at a distance of 2–3 mm from the root. pH will decrease in the rhizosphere
when more cations than anions are absorbed, which occurs, for example, when
plants are given NH_4-N rather then NO_3-N. This is the natural situation with
legumes, where NO_3^- uptake is minimal (Aguilar and van Diest, 1981). Soon
and Miller (1977) found increased P uptake by maize when it was supplied with
NH_4-N as opposed to NO_3-N, due to increased phosphorus availability as a result of a lowering of the rhizosphere pH. Similar effects have been observed for
the uptake of Fe, Zn and Mn (Sarkar and Wyn Jones, 1982). This may be of little
relevance in agricultural practice however, since NH_4^+ is readily oxidised to NO_3^-
in field soils.

Certain plant species respond to nutrient deficiencies by taking up more cations than anions and excreting H^+ ions locally, which solubilize more of the
deficient nutrient. Rape has the ability to do this in phosphate deficient soil (Hedley et al., 1982) as does sunflower when it is deficient in iron and certain varieties
of soybean and corn (Römheld and Marschner, 1984). Nye (1986) has calculated
the pH distribution around roots resulting from the release of H^+ or HCO_3^-, as
well as the potential amount of nutrient absorbed by a root as a result of releasing
a solubilizing agent. As yet these complex and site specific processes have not been
incorporated into the nutrient uptake models discussed in the previous section.

Inorganic and Organic Toxicities

The definition of a toxic substance is not straightforward since even the essential nutrients may have "toxic" effects if they are present in the rooting environment in large quantities. Thus concentrations of Cu and Zn of over 1 ppm in solution inhibit root growth and disrupt plant metabolism, whilst large applications of fertilizer, or saline soil conditions, can cause plasmolysis of root cells. Some plant species are much more tolerant than others to the effects of potentially toxic elements, and several "accumulator" species have been identified (Baker, 1981). For the present purpose elements will be considered toxic if they are non-essential and if small quantities impair plant functions and growth of a wide range of species. The present discussion will be restricted to those toxic elements which may limit the productivity of soils or which pose a threat to human health after entering the food chain. The elements of interest are Al and the heavy metals Pb, Ni, Cr and Cd.

Aluminium is a major constituent of all soils but in neutral and alkaline soils its solubility is too low to cause problems. At pH < 5 appreciable levels of soluble aluminium are present and this, rather than the low pH, reduces plant growth. Root growth is rapidly affected through impaired cell division, short thick roots are produced and branching is inhibited. The symptoms of Al toxicity resemble those of phosphate (and calcium) deficiency but it is not known if this is caused by the precipitation of aluminium phosphates in the soil solution or within root cells. Plant species and cultivars differ in their tolerance to Al, which may be due to its exclusion by the plasmalemma, to root-induced pH changes or to exudation of Al-complexing agents into the rhizosphere (see Foy et al., 1978; Bollard, 1983).

The heavy metals occur naturally in all soils albeit at very low levels. Some soils however contain considerable amounts; for example soils derived from serpentine rocks have naturally high levels of Ni and Cr, and soils and spoil heaps in mining regions may be rich in Cu, Zn or other metals. Heavy metals are being deposited on agricultural soils from a variety of sources, and their build-up there and entry into the food chain in plants is a cause for concern (Bowie and Thornton, 1980). Industrial metal smelting and refining and petrol combustion release metals into the atmosphere which contaminate soils and growing crops, but the immediate concern here is high metal levels on leaf surfaces and other edible plant parts. The main input of heavy metals into agricultural soils is from contaminated sewage sludge (McGrath, 1985).

Heavy metals interact strongly with soils, so their plant availability is generally very low, increasing with decreasing pH and decreasing cation exchange capacity. All are strongly adsorbed by organic and clay colloids and many form insoluble crystalline compounds. For these reasons moderate amounts of contaminated sludge may be applied to neutral soils over a period of years without affecting their productivity (Archer, 1985). Toxic levels in solution are very low (usually less than 1 ppm) and the primary effect is inhibition of root growth. Ni, Cr and Pb tend to remain in the roots, and less than 1% of that absorbed is translocated to the shoots, but Cd is more mobile and resembles Ca in its translocation. Besides affecting root growth heavy metals disrupt cell metabolism, usually by replacing essential metals at physiologically active centres. Plants have developed several mechanisms to limit the uptake of heavy metals; some species appear to be able to exclude them from entering cells, others may inactivate them by complex formation or by incorporating them into cell walls (compartmentation) (see Tinker, 1981). The effect of most organic biocides or heavy metals on soil microbial processes is limited (Domsch, 1984), but recent work has indicated that large dressings of contaminated sewage sludge affect the size of the soil biomass and thereby impair some biological transformations in the soil such as N fixation by blue-green algae (Brookes and McGrath, 1984). The effect of heavy metals in sludges on N fixation by legumes remains to be studied.

Organic toxins in the soil may arise from three sources: plants (released by roots or decaying leaf litter), microbes and their transformations, and applications of agrochemicals (insecticides, fungicides, herbicides). Relatively little conclusive evidence is available on the effect of one plant on the growth of another as a result of the production of harmful chemicals. This subject, allelopathy, has been reviewed by Rice (1974). Roots and soil microbes undoubtedly produce an enormous number of metabolites, some of which are toxic (Kershaw, 1973; Lynch, 1976) but these may be easily destroyed or adsorbed in soil. Specific instances of mutual exclusion of roots as a result of allelopathy are known, thus Israel et al. (1973) attributed the mutual exclusion of peach tree roots to the release of HCN and C_6H_5CHO into the soil. However, the fact that root systems do not inter-penetrate (Raper and Barber, 1970; Baldwin and Tinker, 1972) may indicate simple competition for water and nutrients, because root branching is more extensive if water and nutrients are in plentiful supply.

Organic compounds originating from microbial degradation processes commonly occur in anaerobic soils. Simple aliphatic acids (acetic, butyric, propionic) are formed by the microbial fermentation of cellulose. In topsoils straw is the main substrate whereas in subsoils it is roots. These acids have been found to reduce seedling root growth substantially, when new roots grow within a few centimetres of decomposing straw in soil (Lynch, 1983). Aromatic acids are also formed in soils and are as toxic to roots as aliphatic acids, but there is no clear evidence that they ever reach phytotoxic levels in soil. Ethylene is also produced during anaerobic microbial fermentation and this gas is very effective at inhibiting the extension of cereal roots (Crossett and Campbell, 1975). The effects of ethylene have recently been reviewed by Jackson (1985).

The effects of pesticides on root growth and rhizosphere populations has received little attention so far, and the prevailing opinion is that most of these compounds degrade or are inactivated in the soil. Pesticide transport and degradation in soils has been reviewed by Graham-Bryce (1981).

Conclusion

The interaction of soils and roots is exceedingly complicated. Most of the processes are understood in outline, and their effects are broadly predictable. However, there is still a great deal to be done before all these processes, in particular the microbiological ones, can be integrated into a full and quantitative statement of how a plant grows in soil.

References

Addiscott, T.M., Wagenet, R.J. (1985): Concepts of solute leaching in soils: a review of modelling approaches. Journal of Soil Science *36*:411–425

Aguilar, S.A., van Diest, A. (1981): Rock-phosphate mobilisation induced by the alkaline uptake pattern of legumes utilizing symbiotically fixed nitrogen. Plant and Soil *61*:27–42

Archer, J. (1985): Crop Nutrition and Fertilizer Use. Farming Press: Ipswich, UK

Armstrong, W. (1978): Root aeration in the wetland condition. In: Plant life in anaerobic environments. Hook D., Crawford R.M.M. (eds) Ann Arbor Science Publishers, Ann Arbor, Mich. pp 269–297

Asher, C.J., Edwards, D.G. (1983): Modern Solution Culture Techniques. In: Encyclopedia of Plant Physiology. New Series 15A, Inorganic Plant Nutrition. Läuchli A., Bieleski R.L. (eds) Springer-Verlag: Berlin

Bache, B.W. (1980): In: Effects of Acid Precipitation on Terrestrial Ecosystems. Hutchinson T.C., Havas M. (eds) Plenum: New York, pp 173–182

Baker, A.J.M. (1981): Accumulators and excluders – strategies in the response of plants to heavy metals. Journal of Plant Nutrition *3*:643–654

Baldwin, J.P., Nye, P.H., Tinker, P.B. (1973): Uptake of solutes by multiple root systems from soil. III. A model for calculating the solute uptake by a randomly dispersed root system developing in a finite volume of soil. Plant and Soil *38*:621–635

Baldwin, J.P., Tinker, P.B. (1972): A method for estimating the lengths and spatial pattern of two inter-penetrating root systems. Plant and Soil *37*:209–213

Barber, S.A. (1974): Influence of the plant root on ion movement in soil. In: The Plant Root and its Environment. Carson E.W. (ed) University Press of Virginia: Charlottesville, pp 525–564

Barber, S.A. (1984): Soil Nutrient Bioavailability: A Mechanistic Approach. John Wiley: New York

Barber, S.A., Cushman, J.H. (1981): Nitrogen uptake model for agronomic crops. In: Modelling Waste Water Renovation – Land Treatment. Iskandur I.K. (ed) Wiley-Interscience: New York, pp 382–409

Barraclough, P.B. (1986a): The growth and activity of winter wheat roots in the field: Nutrient uptakes of high-yielding crops. Journal of Agricultural Science, Cambridge *106*:45–52

Barraclough, P.B. (1986b): The growth and activity of winter wheat roots in the field: Nutrient inflows of high-yielding crops. Journal of Agricultural Science, Cambridge *106*:53–59

Bartlett, E.M., Lewis, D.H. (1973): Surface phosphatase activity of mycorrhizal roots of beech. Soil Biology and Biochemistry *5*:249–257

Bhat, K.K.S., Nye, P.H., Baldwin, J.P. (1976): Diffusion of phosphates to plant roots in soil. IV. The concentration-distance profile in the rhizosphere of roots with root hairs in a low P-soil. Plant and Soil *44*:63–72

Bollard, E.G. (1983): Involvement of unusual elements in plant growth and nutrition. In: Encyclopedia of Plant Physiology, New Series 15B, Inorganic Plant Nutrition. Läuchli A., Bieleski R.L. (eds) Springer-Verlag: Berlin

Bowie, S.H.U., Thornton, I. (1985): Environmental Geochemistry and Health. Reidel Publishing Co: Dordrecht

Breeze, V.G., Wild, A. (1984): The uptake of phosphate by plants from flowing nutrient solution. II. Growth of *Lolium Perenne L* at constant phosphate concentrations. Journal of Experimental Botany *35*:1210–1221

Brewster, J.L., Tinker, P.B. (1970): Nutrient cation flow in soil around plant roots. Soil Science Society of America Proceedings *34*:421–426

Brookes, P.C., McGrath, S. (1984): Effects of metal toxicity on the size of the soil biomass. Journal of Soil Science *35*:341–346

Brookes, P.C., Powlson, D.S., Jenkinson, D.S. (1985): The microbial biomass in soil. In: Ecological interactions in soil. Sp. Publ. No. 4 of British Ecological Society. Fitter A.H. (ed) Blackwell: Oxford, pp 123–125

Burns, I.G. (1980): Influence of the spatial distribution of nitrate on the uptake of N by plants: A review and a model for rooting depth. The Journal of Soil Science *31*:155–173

Burns, R.G. (ed) (1978): Soil enzymes. Academic Press: London. pp 380

Caradus, J.R. (1982): Genetic differences in the length of root hairs in white clover and their effect on phosphorus uptake. In: Proceedings of the 9th International Plant Nutrition Colloquium. Warwick, England. Scaife A. (ed) Commonwealth Agricultural Bureaux, pp 84–88

Carson, E.W. (1974): The plant root and its environment. Univ. Virginia Press: Charlottesville, pp 691

Chapin, F.S. (1982): Patterns of phosphorus absorption and chemistry as adaptations to infertile soil. Proceedings of the 9th International Plant Nutrition Colloquium. Warwick, England. Scaife A. (ed) Commonwealth Agricultural Bureaux, Farnham Royal

Claassen, N., Barber, S.A. (1976): Simulation model for nutrient uptake from soil by a growing plant root system. Agronomy Journal 68:961–964

Claassen, N., Jungk, A. (1984): Effect of K uptake rate, root growth and root hairs on potassium uptake efficiency of several plant species. Zeitschrift für Pflanzenernährung und Bodenkunde 147:276–289

Clarkson, D.T., Sanderson, J., Russell, R.S. (1968): Ion uptake and root age. Nature 220:805–806

Clement, C.R., Hopper, M.J., Jones, L.H.P. (1978): The uptake of nitrate by Lolium perenne from flowing nutrient solution, I. Effect of NO_3^- concentration. Journal of Experimental Botany 29:453–464

Cooke, G.W., Williams, R.J.B. (1970): Losses of N and P from agricultural land. Water Treatment and Examination 19:253–276

Cress, W.A., Throneberry, G.O., Lindsey, D.L. (1979): Kinetics of phosphorus absorption by mycorrhizal and non-mycorrhizal tomato roots. Plant Physiology 64:484–487

Crossett, R.N., Campbell, D.J. (1975): The effects of ethylene in the root environment upon the development of barley. Plant and Soil 42:453–464

Domsch, K.H. (1984): Effects of pesticides and heavy metals on biological processes in soil. In: Biological Processes and Soil Fertility. Tinsley J., Darbyshire J.F. (eds) Martinus Nijhoff: Hague, pp 367–378

Drew, M.C. (1975): Comparison of the effects of a localised supply of phosphate, nitrate, ammonium and potassium on the growth of the seminal root system, and the shoot of barley. New Phytologist 75:479–490

Drew, M.C., Saker, L.R. (1978): Nutrient supply and the growth of the seminal root system in barley. III. Compensatory increases in growth of lateral roots, and in rates of phosphate uptake in response to a localised supply of phosphate. Journal of Experimental Botany 29:435–451

Ewens, M., Leigh, R.A. (1985): The effect of nutrient solution composition on the length of root hairs of wheat. Journal of Experimental Botany 36:713–724

Foehse, D., Jungk, A. (1983): Influence of phosphate and nitrate supply on root hair formation of rape, spinach and tomato plants. Plant and Soil 74:359–368

Foy, C.D., Chaney, R.L., White, M.C. (1978): The physiology of metal toxicity in plants. Annual Review of Plant Physiology 29:511–566

Gardner, W.K., Parberry, D.G., Barber, D.A. (1982): The acquisition of phosphorus by Lupinus albus L I. Some characteristics of the soil/root interface. Plant Soil 68:19–32

Gerwitz, A., Page, E.R. (1974): An empirical mathematical model to describe plant root systems. Journal of Applied Ecology 11:773–781

Giller, K.E., Day, J.M. (1985): Nitrogen fixation in the rhizosphere: significance in natural and agricultural systems. In: Ecological interactions in soil. Sp. Publ. No. 4 of British Ecological Society. Fitter A.H. (ed) Blackwell: Oxford, pp 127–147

Glass, A.D., Siddiqi, M.Y. (1984): The control of nutrient uptake rates in relation to the inorganic composition of plants. Advances in Plant Nutrition I. Tinker P.B., Läuchli A. (eds) Praeger Publishers: New York, pp 103–148

Graham-Bryce, I.G. (1981): The behaviour of pesticides in soil. In: The Chemistry of Soil Processes. Greenland D.J., Hayes M.H.B. (eds) John Wiley: Chichester, England, pp 621–671

Gregory, P.J., Crawford, D.V., McGowan, M. (1979): Nutrient relations of winter wheat. 2. Movement of nutrients to the root and their uptake. Journal of Agricultural Science, Cambridge 93:495–504

Hale, M.G., Foy, C.L., Shay, F.J. (1971): Factors affecting root exudation. Advances in Agronomy 23:89–109

Harley, J.L., Smith, S.E. (1983): Mycorrhizal Symbioses. Academic Press: London, pp 483

Hedley, M.J., Nye, P.H., White, R.E. (1982): Plant-induced changes in the rhizosphere of rape seedlings. II. Origin of the pH change. New Phytologist *91*:31–44

Helal, H.M., Sauerbeck, D.R. (1984): Influence of plant roots on C and P metabolism in soil. In: Biological Processes and Soil Fertility. Tinsley J., Darbyshire J.F. (eds) Martinus Nijhoff/ Dr. W. Junk: The Hague, pp 175–182

Hendricks, L., Claassen, N., Jungk, A. (1981): Phosphatverarmung des wurzelnahen Bodens and Phosphataufnahme von Mais und Raps. Zeitschrift für Pflanzenernährung und Bodenkunde *44*:486–499

Israel, D.W., Giddens, J.E., Powell, W.W. (1973): The toxicity of peach tree roots. Plant and Soil *39*:103–112

Itoh, S., Barber, S.A. (1983): Phosphorus uptake by six plant species as related to root hairs. Agronomy Journal *75*:457–461

Jackson, M.B. (1985): Ethylene and responses of plants to soil waterlogging and submergence. Annual Review of Plant Physiology *36*:145–174

Jungk, A., Claassen, N., Kuchenbuch, R. (1982): Potassium depletion of the soil/root interface in relation to soil parameters and root properties. In: Proceedings of the 9th International Plant Nutrition Colloquium. Warwick, England. Scaife A. (ed) Commonwealth Agricultural Bureaux, pp 250–255

Kershaw, K.A. (1973): Quantitative and Dynamic Plant Ecology. Edward Arnold: London

Lethbridge, G., Davidson, M.S. (1983): Root-associated nitrogen fixing bacteria and their role in the nitrogen nutrition of wheat estimated by ^{15}N isotope dilution. Soil Biology and Biochemistry *15*:365–374

Lynch, J.M. (1976): Products of soil micro-organisms in relation to plant growth. CRC Critical Reviews in Microbiology *5*:67–107

Lynch, J.M. (1982): Interactions between bacteria and plants in the root environment. In: Bacteria and plants. Rhodes-Roberts M., Skinner F.A. (eds) Academic Press: London, pp 1–24

Lynch, J.M. (1983): Soil Biotechnology: Microbiological Factors in Crop Productivity. Blackwell: Oxford

Marschner, H. (1983): General introduction to the mineral nutrition of plants. In: Encyclopedia of Plant Physiology. New Series 15A, Inorganic Plant Nutrition. Läuchli A., Bieleski R.L. (eds) Springer-Verlag: Berlin

Mengel, K., Kirkby, E.A. (1982): Principles of Plant Nutrition, 3rd edition. IPI: Berne

McGrath, S.P. (1985): The effects of increasing yields on the macro- and micronutrient concentrations and offtakes in the grain of winter wheat. Journal of the Science of Food and Agriculture *36*:1073–1083

Munns, D.N. (1986): Acid soil tolerance in legumes and Rhizobia. Advances in Plant Nutrition *2*:63–92

Newman, E.I. (1985): The rhizosphere: carbon sources and microbial populations. In: Ecological interactions in soil. Sp. Publ. No. 4 of British Ecological Society. Fitter A.H. (ed) Blackwell: Oxford, pp 107–121

Newman, E.I., Watson, A. (1977): Microbial abundance in the rhizosphere: a computer model. Plant and Soil *48*:17–56

Nye, P.H. (1979): Diffusion of ions and uncharged solutes in soils and soil clays. Advances in Agronomy *31*:225–272

Nye, P.H. (1984): On estimating the uptake of nutrients solubilised near roots or other surfaces. Journal of Soil Science *35*:439–446

Nye, P.H. (1986): Acid-base changes in the rhizosphere. Advances in Plant Nutrition II, Praeger Scientific: New York, pp 129–154

Nye, P.H., Tinker, P.B. (1977): Solute movement in the soil-root system. Blackwell: Oxford

Poljakoff-Mayber, A., Gale, J. (1975): Plants in saline environments. Ecological studies, Vol. 15 Springer-Verlag: Berlin

Ratnayake, M., Leonard, R.T., Menge, J.A. (1978): Root exudation in relation to supply of phosphorus and its possible relevance to mycorrhiza formation. New Phytologist *81*:543–552

Rice, E.L. (1974): Allelopathy. Academic Press: New York

Riley, D., Barber, S.A. (1970): Salt accumulation at the soyabean root-soil interface. Soil Science Society of America Proceedings 34:154–155

Römheld, V., Marschner, H. (1979): Fine regulation of iron uptake by Fe-efficient plant – Helianthus annus. In: The soil root interface. Harley J.L., Russell R.S. (eds) Academic Press: London, pp 405–417

Römheld, V., Marschner, H. (1984): Plant-induced pH changes in the rhizosphere of "Fe-efficient" and "Fe-inefficient" soybean and corn cultivars. Journal of Plant Nutrition 7:623–630

Römheld, V., Marschner, H. (1986): Mobilisation of iron in the rhizosphere of different plant species. Advances in Plant Nutrition II, Praeger Scientific: New York, pp 155–204

Rorison, I. (1969): Ecological inferences from laboratory experiments on mineral nutrition. In: Ecological aspects of the mineral nutrition of plants. Rorison I.H. (ed) Blackwell: Oxford

Rovira, A.D. (1979): Biology of the soil-root interface. In: The soil-root interface. Harley J.L., Russell R.S. (eds) Academic Press: London

Russell, R.S. (1977): Plant root systems. McGraw-Hill: London

Sanders, F.E., Tinker, P.B. (1973): Phosphate flow into mycorrhizal roots. Pesticide Science 4:385–395

Sarkar, A.N., Wyn Jones, R.G. (1982): Effect of rhizosphere pH on availability and uptake of Fe, Mn and Zn. Plant and Soil 66:361–372

Schenk, M.K., Barber, S.A. (1980): Potassium and phosphorus uptake by corn genotypes grown in the field as influenced by root characteristics. Plant and Soil 54:65–76

Schippers, B., Gams, W. (1979): Soil-borne plant pathogens. Academic Press: London

Sharpley, A.N., Reed, L.W. (1982): Effect of environmental stress on the growth and amounts and forms of phosphorus in plants. Agronomy Journal 74:19–22

Silberbush, M., Barber, S.A. (1983): Sensitivity of simulated phosphorus uptake to parameters used by a mechanistic mathematical model. Plant and Soil 74:93–100

Silberbush, M., Barber, S.A. (1984): Phosphorus and potassium uptake of field-grown soybean cultivars predicted by a simulation model. Soil Science Society America Journal 48:592–596

Slankis, V. (1983): Hormonal relationships in mycorrhizal development. In: Ectomycorrhizas. Marks G.C, Kozlowski T.T. (eds) Academic Press: New York

Soon, Y.K., Miller, M.H. (1977): Changes in the rhizosphere due to NH_4^+ and NO_3^- fertilization and phosphorus uptake by corn seedlings (Zea Mays L). Soil Science Society of America Journal 41:77–80

Staunton, S., Nye, P.H. (1983): The self diffusion of Na in a naturally aggregated soil. Journal of Soil Science 34:263–269

Stefanson, R.C. (1973): Evolution patterns of nitrous oxide and nitrogen in sealed soil-plant systems. Soil Biology and Biochemistry 5:167–169

Strebel, O., Grimme, H., Renger, M., Fleige, H. (1980): A field study with ^{15}N of soil and fertilizer nitrate uptake and of water withdrawal by spring wheat. Soil Science 130:205–210

Tinker, P.B. (1975): Effects of vesicular-arbuscular mycorrhizas on higher plants. Symp. Society Experimental Biology 29:325–349

Tinker, P.B. (1976): Roots and water. Phil. Trans. Royal Society London B 273:445–461

Tinker, P.B. (1980): Root-soil interactions in crop plants. In: Critical Reports on Applied Chemistry, vol 2 Soils and Agriculture. Tinker P.B. (ed) SCI/Blackwell: Oxford

Tinker, P.B. (1981): Levels, distribution and chemical forms of trace elements in food plants. Phil. Trans. Royal Society, London B 294:41–55

Tinker, P.B. (1986): Trace elements in arable agriculture. Journal Soil Science 37:587–601

Tinker, P.B., Gildon, A. (1982): Mycorrhizal fungi and ion uptake. In: Metals and micronutrients: Uptake and utilisation by plants. Robb D.A., Pierpoint S. (eds) Academic Press: London, pp 21–32

Trolldenier, G. (1979): The effects of mineral nutrition of plants and soil oxygen on rhizosphere organisms. In: Soil-borne plant pathogens. Shippers B., Gams W. (eds) Academic Press: London, p 235

Wehrmann, J., Scharpf, H.C., Böhmer, M., Wollring, J. (1982): Determination of nitrogen fertilizer requirements by nitrate analysis of the soil and of the plant. Proceedings of the 9th International Plant Nutrition Colloquium: Warwick, England. Scaife A. (ed) Commonwealth Agricultural Bureaux, pp 702–709

Whipps, J.M., Lynch, J.M. (1985): Energy losses by the plant in rhizodeposition. Sum Proceedings Phytochemical Society Europe 26:59–71

Wild, A., Breeze, V.G. (1981): Nutrient uptake in relation to growth. In: Physiological Processes Limiting Plant Productivity. Johnson C.B. (ed) Butterworths: London

Willigen, P. de (1984): Some theoretical aspects of the influence of soil-root contact on uptake and transport of nutrients and water. ILRI Publication No. 37, pp 268–275

Reaction Types in the Environment

C. M. Menzie

American University
Washington, D.C., USA

The environment in which we live exposes man and all other animals to chemicals that have been elaborated by other life forms, released to control other life forms, given to cure or prevent disease or affect growth, or propagated as a result of man's technology. These compounds may appear in foods, medicines, solvents, detergents, cosmetics, the atmosphere, and in our water supplies; and they may interact with enzymes, substrates, or chemical impulse transmitters in such ways as to affect and influence behaviour.

It has been estimated that about one-third of all industrial chemicals affect CNS with functional changes preceding obvious morphology injury. Understanding the biotransformations of chemicals and their metabolic pathways would help us to understand these changes and the influence of chemicals on behaviour and biochemical lesions. We know that the metabolism and biotransformation of many compounds lead to more polar material, decreased lipid solubility, and increased elimination. We know too that, while hydroxylation generally may be the first step in deactivation, N-hydroxylation can be an activation step that leads to conjugates capable of non-enzymatic reaction with tissue macromolecules under physiological conditions.

In Vol. 2/Part A, reaction types were set forth and their ubiquitous nature was indicated. In the following, the generality in nature of these reactions is further demonstrated.

	Compound	Metabolite	Reference
A. Ring Hydroxylation			
1. Agricultural Products			
a) Mammal			
Sheep	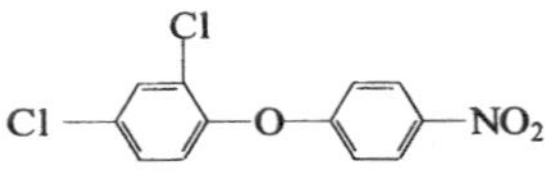	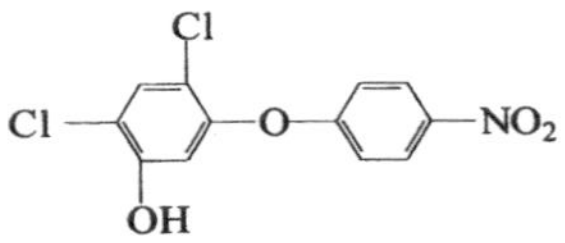	Hunt et al. [45]
Rat, Pig	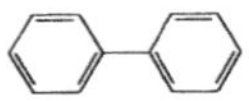	2-Hydroxybiphenyl 3-Hydroxybiphenyl 4-Hydroxybiphenyl	Meyer and Scheline [87]; Meyer et al. [86]
Rabbit	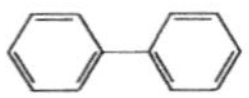	2-Hydroxybiphenyl 4-Hydroxybiphenyl 4,4'-Dihydroxybiphenyl	Berninger et al. [10]; Meyer [85]
b) Bird			
Chicken	Endrin	Anti-12-hydroxyendrin	Baldwin et al. [9]
c) Fish and Shellfish			
Salvelinus fontinalis, Raja ocellata, Homarus americus, Mytilus edulis, Asterias vulgares, and *Cancer irroratus*	Biphenyl	2-Hydroxybiphenyl 4-Hydroxybiphenyl	Willis and Addison [137]
Mosquitofish	Aldrin	9-Hydroxydieldrin	Metcalf et al. [84]
d) Insect			
American cockroach	Endrin	2-Hydroxyendrin	Kadous [54]
Saturniid larvae	Endrin	Hydroxylated endrin	Krieger et al. [66]
e) Plant			
Soybean, rice, wheat, peanuts, *Senecio* sp., and *Chenopodium* sp.	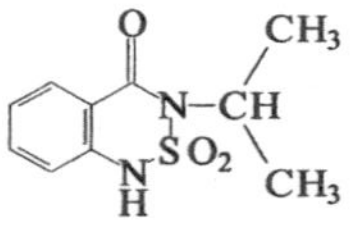	6-Hydroxybentazon 8-Hydroxybentazon	Cannizzaro [15]; Mine et al. [89]; Otto [107]

Carrots and cotton

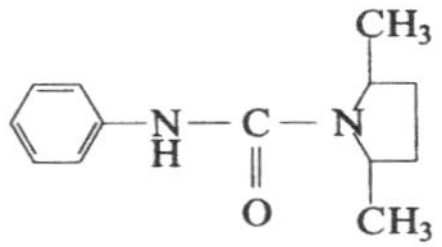

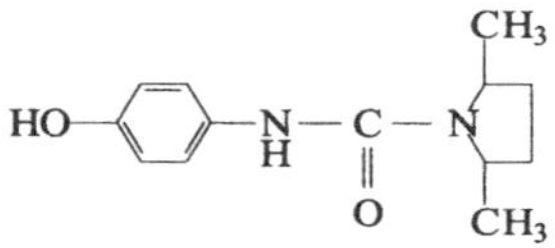

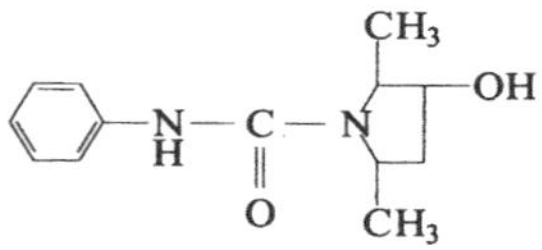

Frear and Swanson [33];
Frear et al. [34]

Western bracken

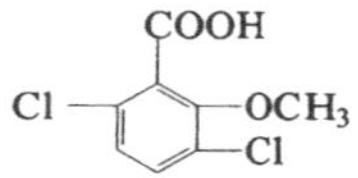

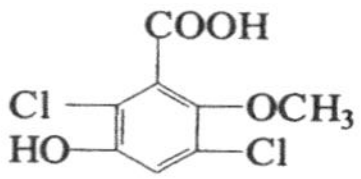

Robocker and Zamora [116]

Bentazon
Biphenyl

Hydroxybentazon
2-Hydroxybiphenyl
4-Hydroxybiphenyl

f) Mircoorganisms
Soil microorganisms
Zooplankton

Kupelian [68]
Willis and Addison [137]

2. Industrial Products

a) Mammal
Rat

Mate et al. [80], Tulp and
Hutzinger [131]

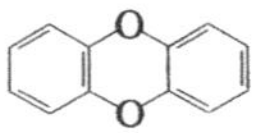

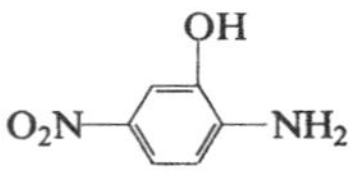

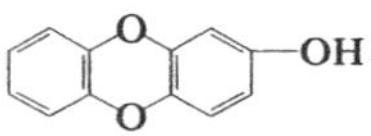

Dog tissue slices

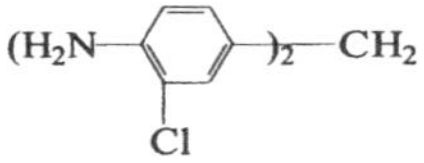

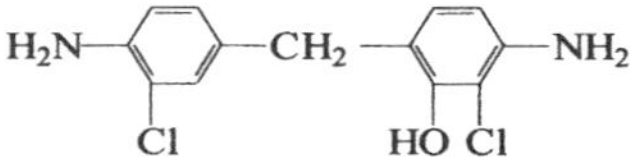

Manis and Braselton [78]

180 C. M. Menzie

	Compound	Metabolite	Reference
b) Bird			
Duckling and chick hepatic supernate	Aflatoxin B$_1$	Aflatoxin hemiacetal	Patterson and Roberts [109]
Red-winged blackbird liver and kestrel liver	Aniline	p-hydroxyaniline	Pan et al. [108]; Wit [139]
c) Microorganism			
Fusarium oxysporum	2-Chloroaniline 3-Chloroaniline 4-Chloroaniline	o-hydroxyaniline analogs	Fletcher and Kaufman [29]
Sewage sludge	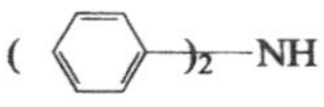	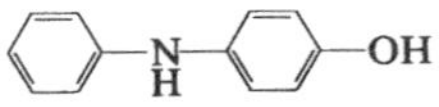	Gardner et al. [35]

B. Ether Cleavage
1. Agricultural Products

	Compound	Metabolite	Reference
a) Mammal			
Sheep	Nitrofen	2,4-dichlorophenol	Hunt et al. [45]
Rat	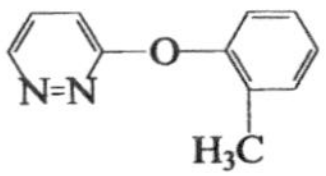	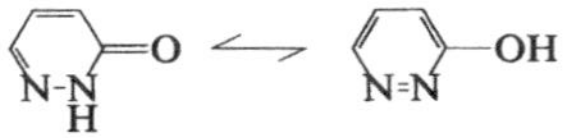	Nakagawa and Ando [95]
Dog	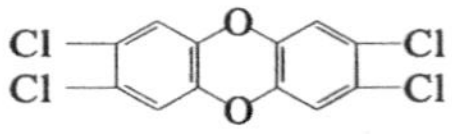	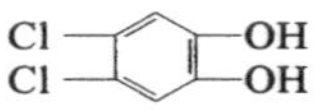	Poiger et al. [110]
b) Fish			
Pseudorasbora parva	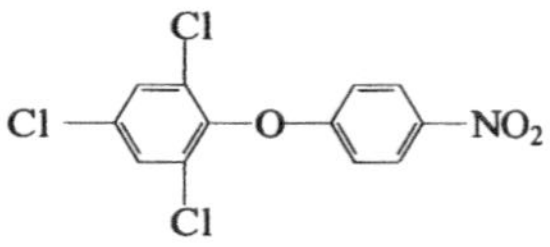	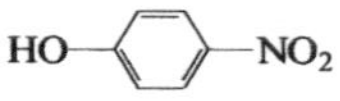	Kanazawa and Tomizawa [55]

c) Plant			
Rice seedling and barnyard millet	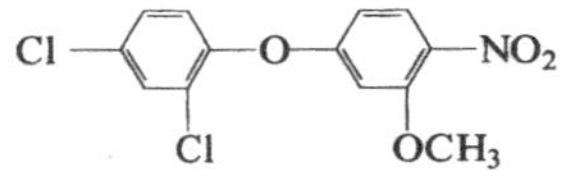	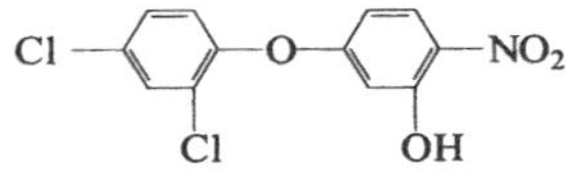	Niki et al. [101]
Barley and tomato	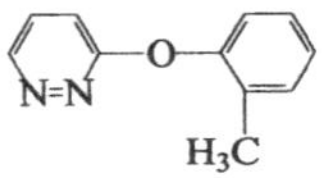	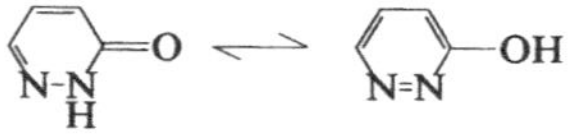	Nakagawa et al. [96]
d) Microorganism			
2 bacteria, a fungus and an actinomycete (all not further ident.)	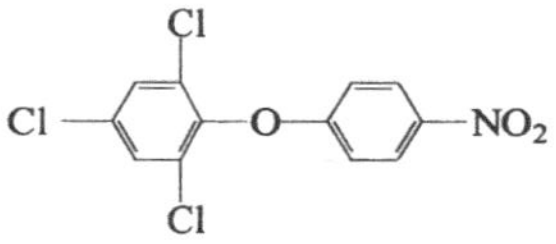	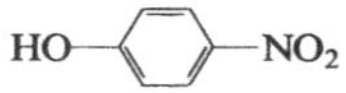	Kuwatsuka et al. [70]
In flooded soil	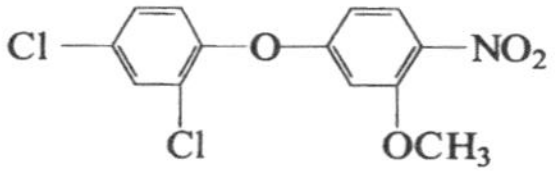	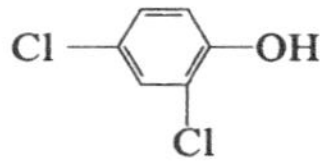	Niki and Kuwatsuka [100]
2. Industrial Products			
a) Mammal			
Rat	$Cl{-}CH_2{-}O{-}CH_3$	CH_3OH	Van Duuren [144]
b) Bird			
Red-winged blackbird liver and kestrel liver	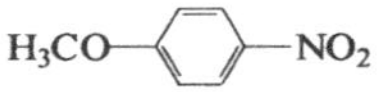	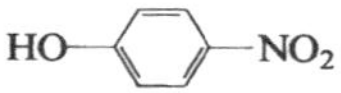	Pan et al. [108]; Wit [139]
c) Fish			
Rainbow trout liver prep.	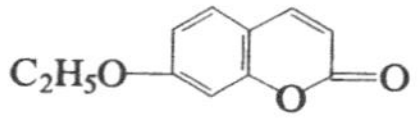	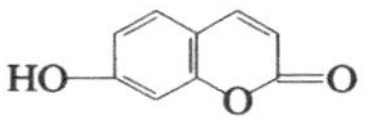	Snegaroff [120]
d) Insect			
Houseflies	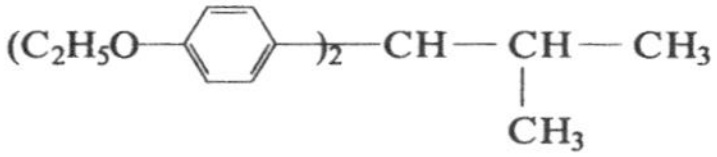	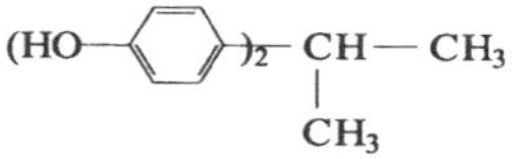	Farooqui and Metcalf [27]

Compound	Metabolite	Reference

C. Sulfoxide/Sulfone Formation

1. Agricultural Products

a) Mammal

Compound	Metabolite	Reference
Cow and rat — Bay NTN 9306	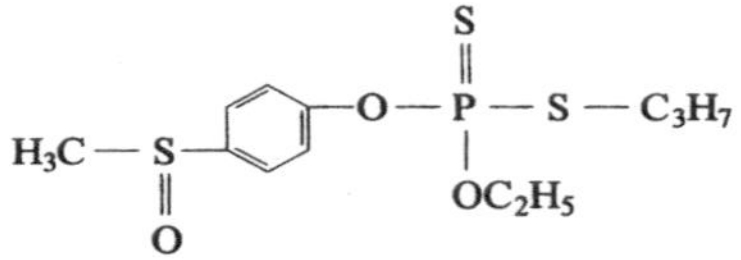 $H_3C-S(=O)-$ [phenyl] $-O-P(=S)(S-C_3H_7)(OC_2H_5)$	Ivie et al. [51]; Bull and Ivie [14]
Rat — Bithional	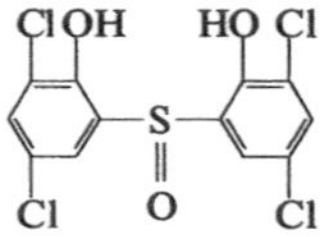 [chlorinated diphenyl sulfoxide structure: HO, Cl; $ClOH$, Cl rings linked by $S=O$] & Sulfone	Meshi et al. [83]
Rat — $O=C(-O-\text{[phenyl-CH}_2]\,)(-N(H)(CH_3))$, $CH_2-S-C_2H_5$	$O=C(-O-\text{[phenyl-CH}_2]\,)(-N(H)(CH_3))$, $CH_2-S(=O)-C_2H_5$	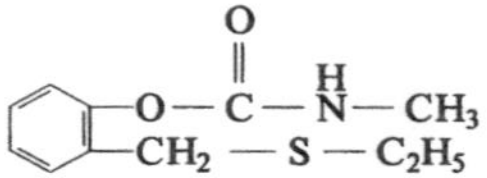 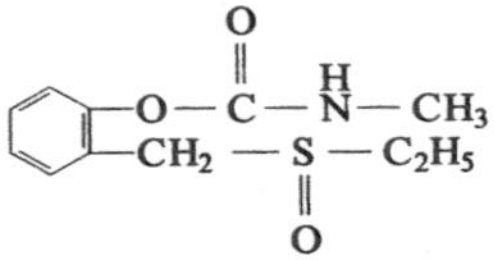 Nye et al. [102]

b) Bird

Compound	Metabolite	Reference
Chicken — Aldicarb	Aldicarb sulfoxide / Aldicarb sulfone	Hicks et al. [42]

c) Invertebrate

Compound	Metabolite	Reference
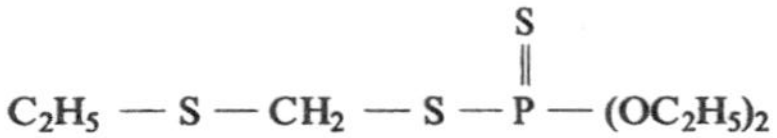 *Panagrellus redivivus* — $C_2H_5-S-CH_2-S-P(=S)-(OC_2H_5)_2$	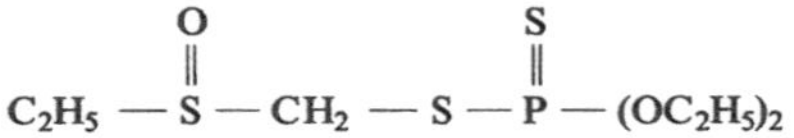 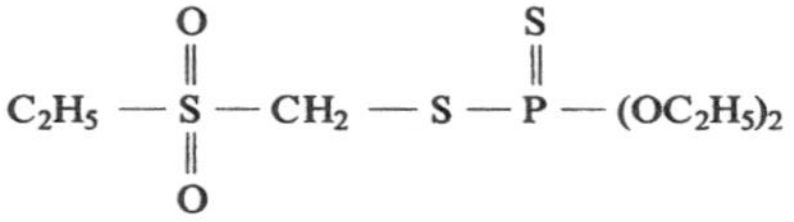 $C_2H_5-S-CH_2-S(=O)-P(=S)-(OC_2H_5)_2$; $C_2H_5-S(=O)-CH_2-S(=O)-P(=S)-(OC_2H_5)_2$	LePatourel and Wright [73]

Tobacco budworm
3rd stage larvae

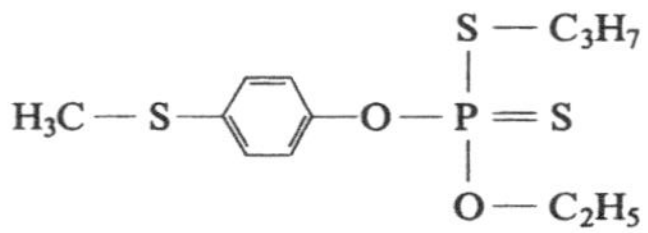

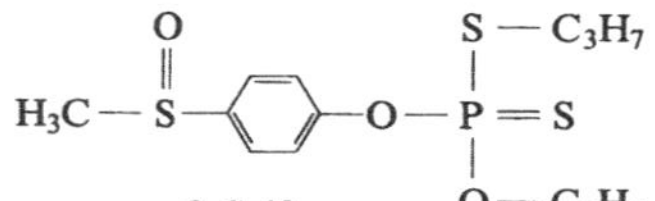

$H_3C-S-\langle\text{aryl}\rangle-O-P(=S)(S-C_3H_7)(O-C_2H_5)$ → $H_3C-S(=O)-\langle\text{aryl}\rangle-O-P(=S)(S-C_3H_7)(O-C_2H_5)$ & Sulfone

Bull [13]

d) Plant

Cabbage seedling and corn

$C_2H_5-S-CH_2-S-P(=S)-(OC_2H_5)_2$ → $C_2H_5-S(=O)-CH_2-S-P(=S)-(OC_2H_5)_2$

Krause and Van Dyk [65];
Lichtenstein et al. [74]

Cotton

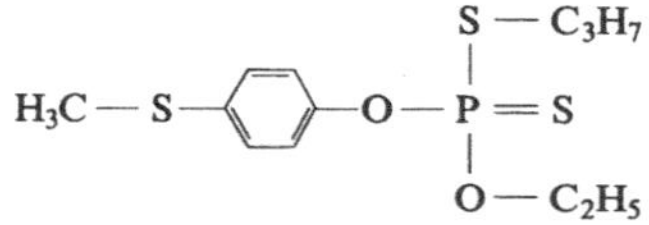

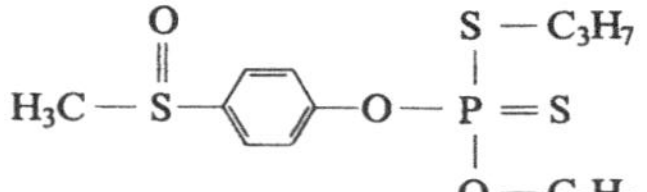

$H_3C-S-\langle\text{aryl}\rangle-O-P(=S)(S-C_3H_7)(O-C_2H_5)$ → $H_3C-S(=O)-\langle\text{aryl}\rangle-O-P(=S)(S-C_3H_7)(O-C_2H_5)$ & Sulfone

Bull and Ivie [14]

Bean

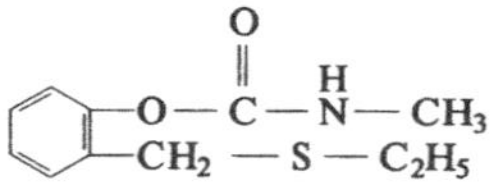

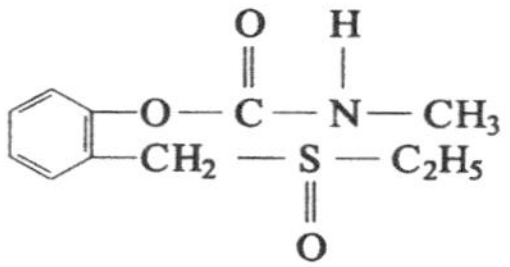

$\langle\text{aryl}\rangle-O-C(=O)-N(H)-CH_3$, $CH_2-S-C_2H_5$ → $\langle\text{aryl}\rangle-O-C(=O)-N(H)-CH_3$, $CH_2-S(=O)-C_2H_5$

Marshall and Dorough [79]

e) Microorganism
In soil

$C_2H_5-S-CH_2-S-P(=S)-(OC_2H_5)_2$ → $C_2H_5-S(=O)-CH_2-S-P(=S)-(OC_2H_5)_2$ → $C_2H_5-S(=O)_2-CH_2-S-P(=S)-(OC_2H_5)_2$

Walter-Echols and Lichtenstein [133–135]; Drager [22]

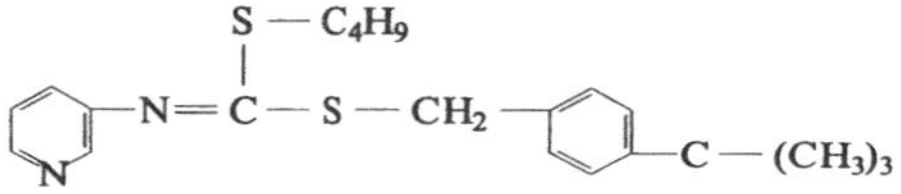

$\langle\text{pyridyl}\rangle-N=C(S-C_4H_9)-S-CH_2-\langle\text{aryl}\rangle-C-(CH_3)_3$ a Sulfoxide

Ohkawa et al. [104]

Compound	Metabolite	Reference

2. Industrial Products

a) Mammal

Rat

CH_3-CH_2 CH_3-CH_2 S (compound) → CH_3-CH_2 CH_3-CH_2 S=O (metabolite)

Hoodi and Damani [44]

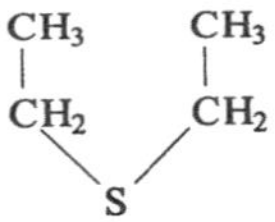

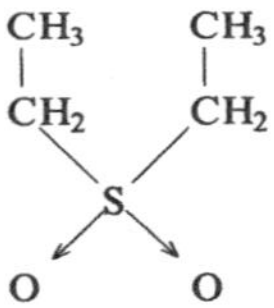

D. Dehydrogenation/Dehalogenation/Dehydrohalogenation

1. Agricultural Products

a) Mammal

Human

Cl Cl (Cl) O—P=S—(OC$_2$H$_5$)$_2$ → (Cl) Cl (Cl) O—P=S—(OC$_2$H$_5$)$_2$

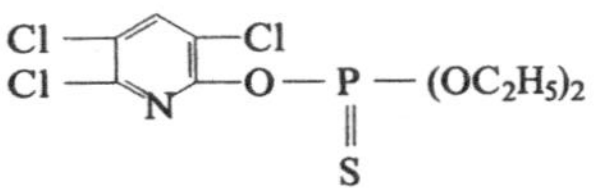

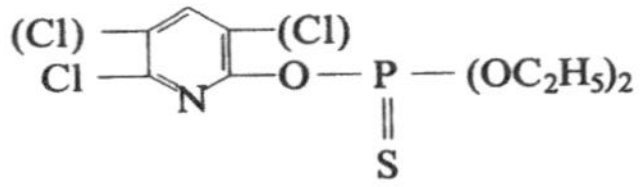

Lores et al. [75, 76]

In Meat fat
Rhesus monkey
Rat

Pentachloronitrobenzene
Pentachlorophenol

(3- or 5-dechlorinated) 5-dechlorinated analog
a Tetrachloroaniline
Tetra- and trichloro-p-hydroquinone

Luke and Dahl [77]
Avrahami and White [8]
Ahlborg [2]; Ahlborg et al. [3]

b) Fish

Carp and rainbow trout

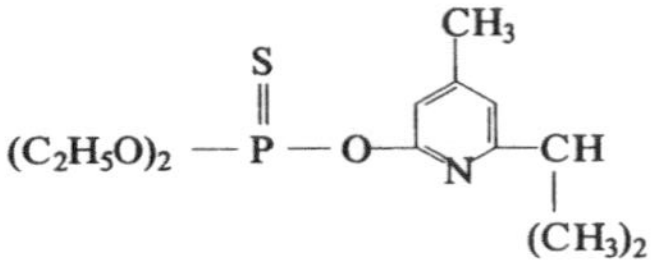

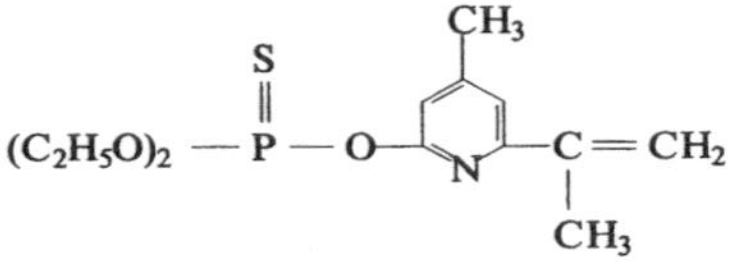

$(C_2H_5O)_2$—P=S—O—...CH$_3$...CH(CH$_3$)$_2$ → $(C_2H_5O)_2$—P=S—O—...CH$_3$...C=CH$_2$ / CH$_3$

Seguchi and Asaka [119]

c) Insect
Eastern subterranean termite

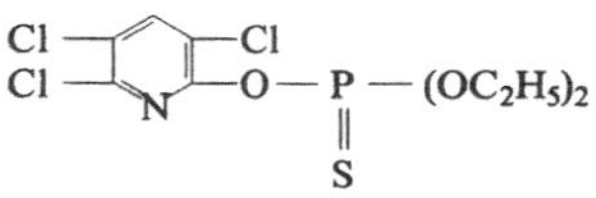

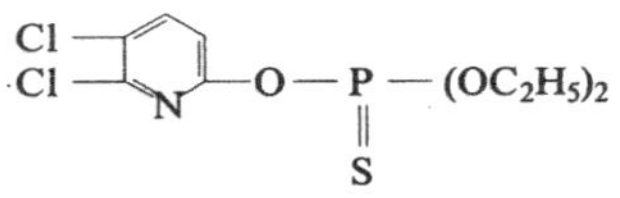

Hutcharern [46]; Hutcharern and Knowles [47]

d) Plant
Onion

Pentachloronitrobenzene → a Tetrachloronitrobenzene

Muller et al. [94]

e) Microorganism
Flooded soil

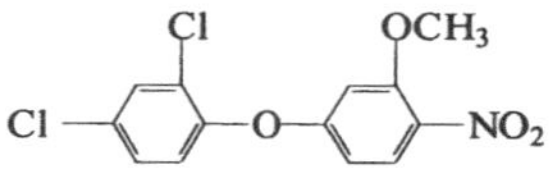

a monochloro analog

Niki and Kuwatsuka [100]

Rice fields and upland fields

Pentachlorophenol → Tetra- and trichlorophenols

Kuwatsuka and Igarashi [69]

2. Industrial Products

a) Mammal
Rat liver prep.

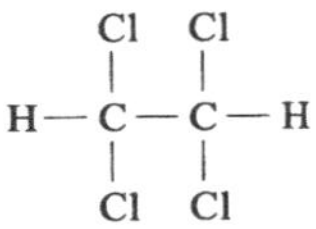

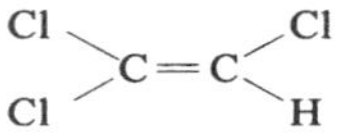

Koizumi et al. [62]

b) Bird
Quail and chickens microsomes

Monocrotaline → Di-dehydromonocrotaline

White et al. [136]

c) Microorganism
From lake sediment and sewage

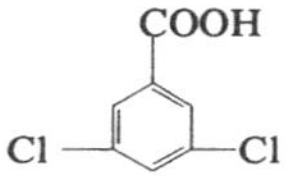

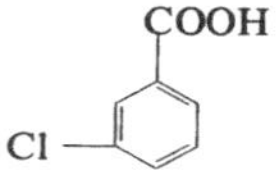

Suflita et al. [124]

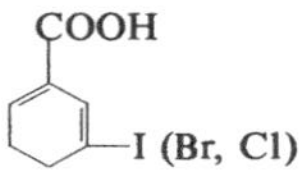

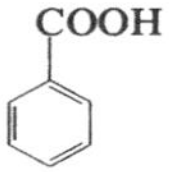

Compound	Metabolite	Reference

E. Reduction

1. Agricultural Products

a) Mammal

	Compound	Metabolite	Reference
Sheep	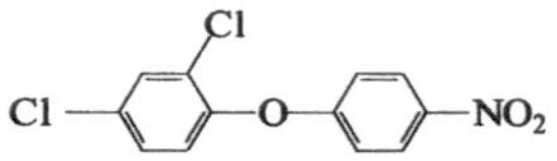 (chlorophenyl)–O–(phenyl)–NO_2 structure (Cl, Cl, O, NO_2)	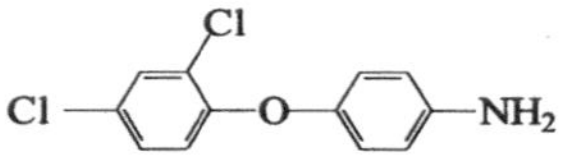 (chlorophenyl)–O–(phenyl)–NH_2 structure (Cl, Cl, O, NH_2)	Hunt et al. [45]
Rat liver microsomes	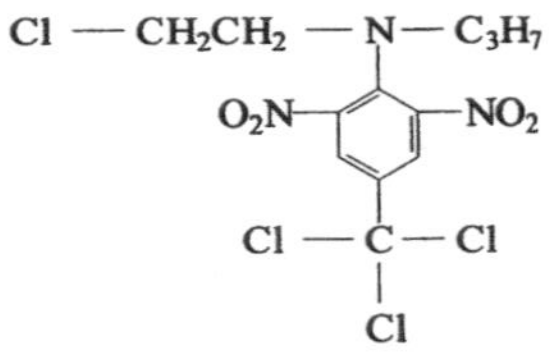 $Cl-CH_2CH_2-N-C_3H_7$; ring with NO_2, O_2N ; $Cl-C-Cl$, Cl	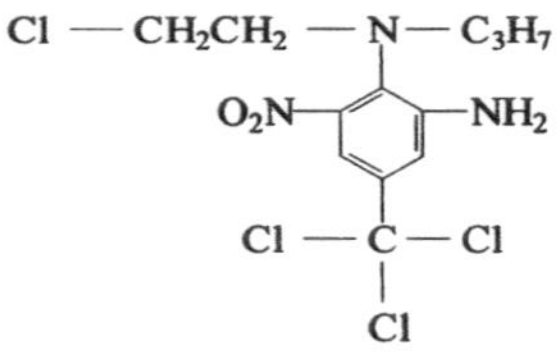 $Cl-CH_2CH_2-N-C_3H_7$; ring with NH_2, O_2N ; $Cl-C-Cl$, Cl	Nelson et al. [97, 98]
Rhesus monkey, goat, rabbit	Pentachloronitrobenzene	Pentachloroaniline	Avrahami and White [8]; Aschbacher and Feil [6]; Renner et al. [113]

b) Bird

	Compound	Metabolite	Reference
White leghorn cockrel	Pentachloronitrobenzene	Pentachloroaniline	Dunn et al. [23]; Reed et al. [112]
Chicken	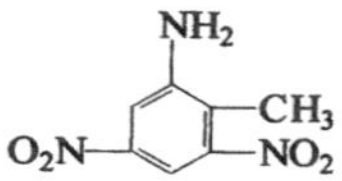 ring with NH_2, CH_3, NO_2, O_2N	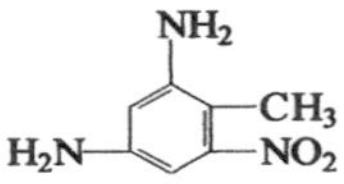 ring with NH_2, CH_3, NO_2, H_2N	Smith et al. [145]

c) Fish

	Compound	Metabolite	Reference
Pseudorasbora parva	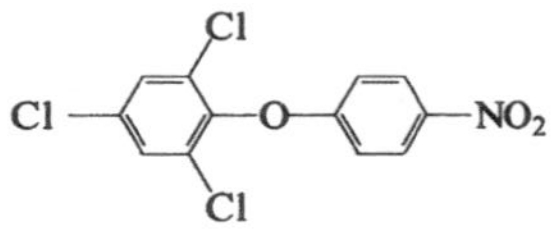 (trichlorophenyl)–O–(phenyl)–NO_2 structure (Cl, Cl, O, Cl, NO_2)	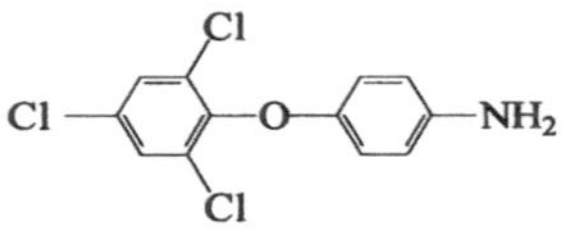 (trichlorophenyl)–O–(phenyl)–NH_2 structure (Cl, Cl, O, Cl, NH_2)	Kanazawa and Tomizawa [55]

	Substrate	Product	Reference
(Not identified)	Bayer 2353	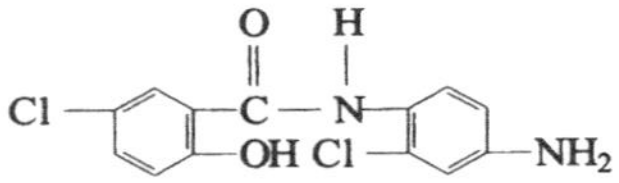 Bayer 2353 (reduced, –NO₂ → –NH₂)	Allen et al. [5]
d) Invertebrate *Ascaris lumbricoides*	Bayer 2353 (Cl–ring–$C(=O)$–$N(H)$–ring(OH, Cl)–NO_2)	Cl–ring–$C(=O)$–$N(H)$–ring(OH, Cl)–NH_2	Douch and Gahagan [21]
e) Plant Onion, peanut	Pentachloronitrobenzene	Pentachloroaniline	Muller et al. [94]; Lamoureux and Rusness [71]
f) Microorganism Flooded soil, bacteria, fungus, actinomycete	Nitrofen, CFNP, CNP, Chlomethoxynil	Respective 4-aminophenyl analogs	Kuwatsuka et al. [70]; Niki and Kuwatsuka [99, 100]
Pseudomonas fragi BK₉	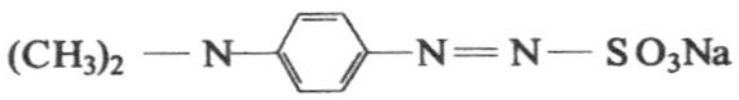 $(CH_3)_2$—N—⟨ring⟩—N=N—SO_3Na	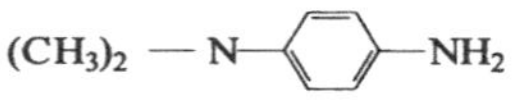 $(CH_3)_2$—N—⟨ring⟩—NH_2	Karanth et al. [58]
2. Industrial Products			
a) Mammal Rat	NH_2—⟨ring⟩—NO_2	NH_2—⟨ring⟩—NH_2	Mate et al. [80]
Rate, mouse, guinea pig	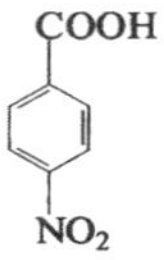 $COOH$—⟨ring⟩—NO_2	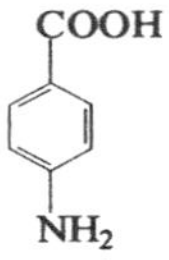 $COOH$—⟨ring⟩—NH_2	Adamson et al. [1]

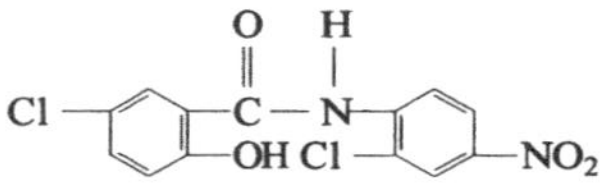

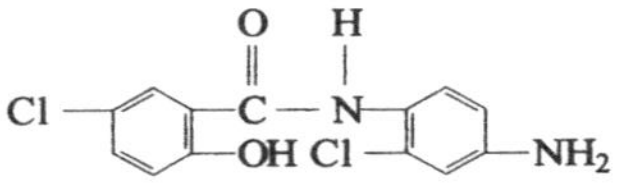

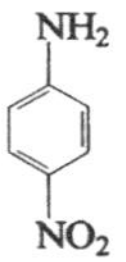

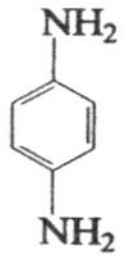

Compound	Metabolite	Reference
Rabbit liver prep.		
Azo dyes (Orange III, Orange IV, Scarlet R)	Respective aniline products	Fouts et al. [32]
b) Bird		
Pigeon liver prep.		
(4-nitrobenzoic acid: COOH––NO_2)	(4-aminobenzoic acid: COOH––NH_2)	Adamson et al. [1]
Kestrel liver		
(4-nitrobenzoic acid: COOH––NO_2)	(4-aminobenzoic acid: COOH––NH_2)	Wit [139]
c) Fish		
Barracuda, yellow-tail snapper		
(4-nitrobenzoic acid: COOH––NO_2)	(4-aminobenzoic acid: COOH––NH_2)	Adamson et al. [1]
d) Reptile		
Turtle, horned toad (liver prep's)		
(4-nitrobenzoic acid: COOH––NO_2)	(4-aminobenzoic acid: COOH––NH_2)	Adamson et al. [1]
e) Microorganism		
Bacterial suspension from human, monkey, and rat feces and by intestinal bacteria		
Benzidine, dimethylbenzidine, dimethoxybenzidine	Respective amines	Cerniglia et al. [146]

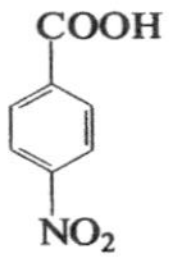

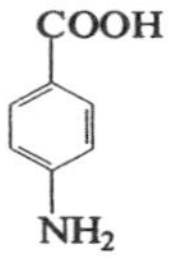

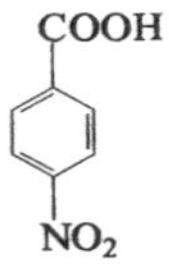

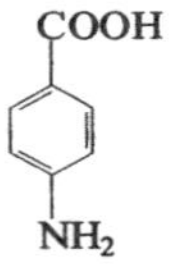

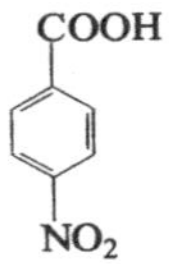

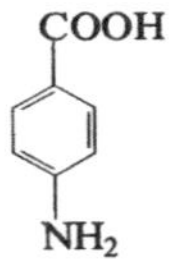

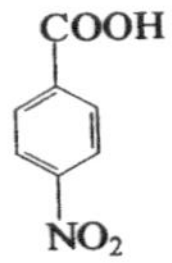

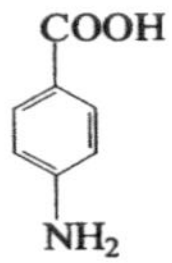

F. Conjugation

1. Agricultural Products

a) Mammal

Rat — Glucuronide — Gaughan et al. [147, 148]

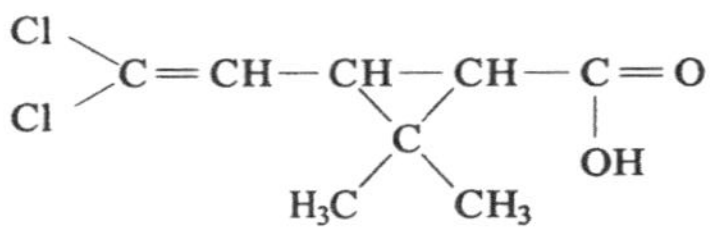

Cow — Glucuronide and Glutamate — Gaughan et al. [147, 121]

Structure (from permethrin):

$Cl_2C=CH-CH-CH-C=O$ with $-OH$, $C(CH_3)_2$, H_3C

b) Bird

Japaneses quail and chicken — OSO_3H substituted (with CH_3, NO_2) — Mihara et al. [88]

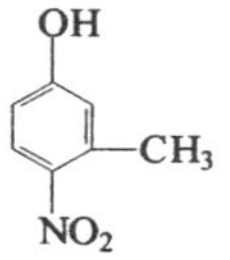

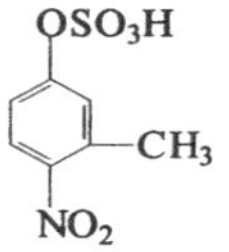

Structure: OH substituted aromatic (CH_3, NO_2)

Chickens — Glucuronide and taurine — Gaughan et al. [149]

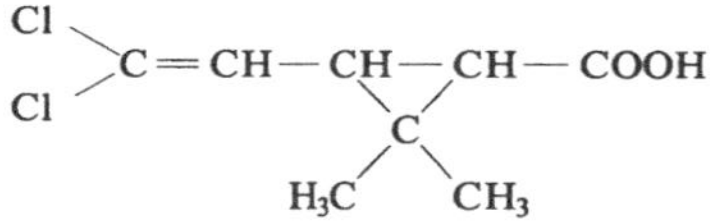

Structure: $Cl_2C=CH-CH-CH-COOH$ with $C(CH_3)_2$, H_3C

c) Fish and Shellfish

Rainbow trout — O—Glucuronide (aromatic with CH_3, NO_2) — Takimoto and Miyamoto [130]

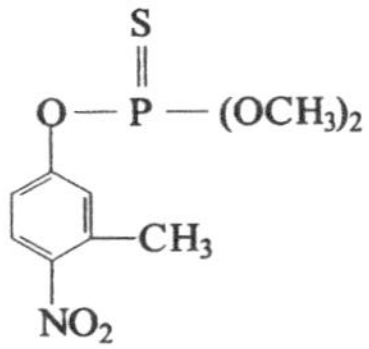

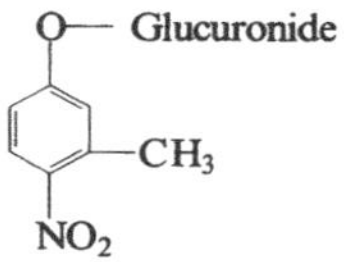

Methyl parathion: $S=P-(OCH_3)_2$, O— aromatic (CH_3, NO_2)

Ricefield crayfish (*Procambarus clarkii*)
Malaysian prawn (*Macrobracium rosenbergii*)
Ridgeback pawn (*Sicyonia ingentis*) — p-Nitrophenylglucopyranoside — Foster and Crosby [31]

Compound	Metabolite	Reference

d) Insect

Housefly

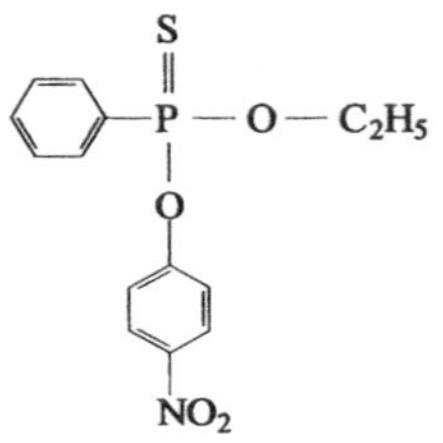

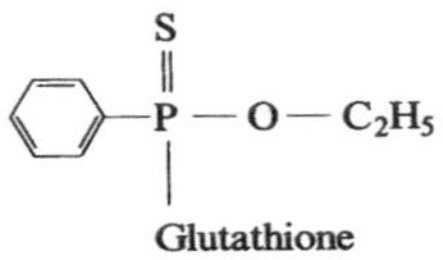

Compound: phenyl–P(=S)(–O–C_2H_5)–O–(C6H4)–NO_2

Metabolite: phenyl–P(=S)(–O–C_2H_5)–Glutathione — Hajjar et al. [38]

e) Plant

Rice

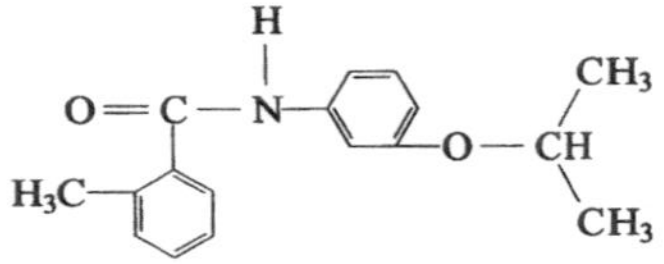

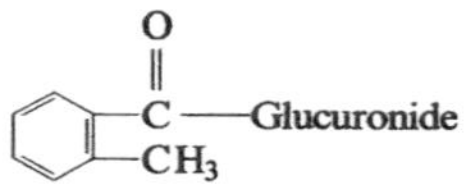

Compound: H_3C–(C6H4)–C(=O)–N(H)–... with –O–CH(–CH_3)(–CH_3)

Metabolite: C(=O)(–CH_3)–Glucuronide — Yumita et al. [143]

2. Industrial Products

a) Mammal

Dog tissue slices

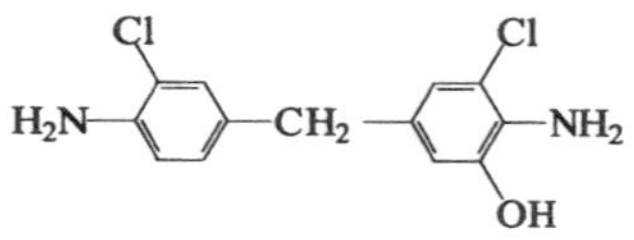

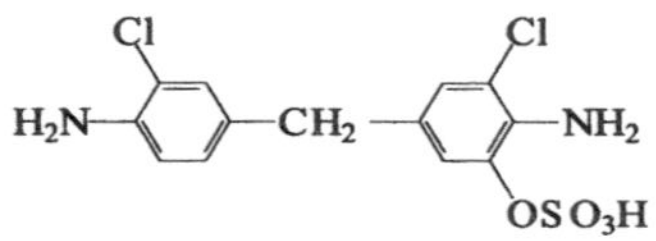

Compound: Cl, NH_2, OH substituted ring; H_2N–(ring)–CH_2–Cl

Metabolite: Cl, NH_2, OSO_3H substituted ring; H_2N–(ring)–CH_2–Cl — Manis and Braselton [78]

b) Bird

Pigeon

Compound: Indoleacetic acid

Metabolite: (indole)–CH_2–C(=O)–NH–CH_2–CH_2–SO_3H — Bridges et al. [12]

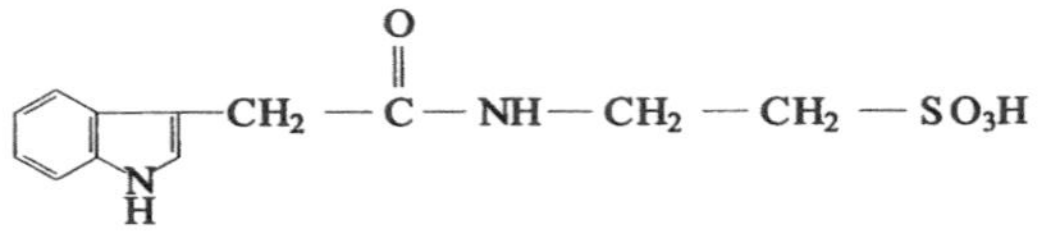

Chicken

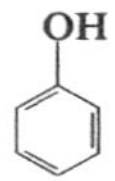

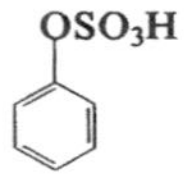

Compound: phenol (OH on ring)

Metabolite: OSO_3H on ring — Capel et al. [16]; Sperber [122]

c) Fish and Shellfish

Spiny lobster

C_6H_5—CH_2—$C(=O)$—taurine → Phenylacetic acid — James [53]

English sole

Respective glucuronides → 2-Naphthol, 9-HO-fluorene, and 1,2-dihydro-1,2-dihydroxyphenanthrene — Krahn et al. [63]

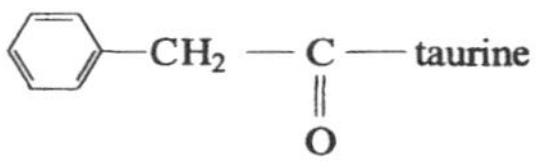

d) Plant

Pine seedlings

naphthyl—CH_2COOH → naphthyl—$CH_2C(=O)$—Aspartic acid — Riov et al. [115]

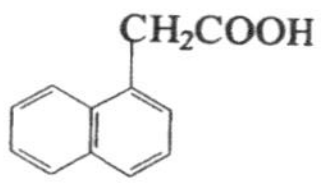

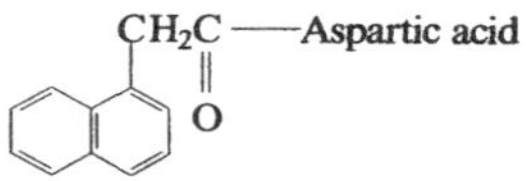

G. Hydrolysis

1. Agricultural Products

a) Mammal

Rat

Cl—C_6H_4—S—CH_2—S—$P(=S)(OC_2H_5)_2$ → $(C_2H_5O)_2$—$P(=S)$—OH, Cl—C_6H_4—SH — DeBaun and Menn [20]; Menn et al. [82]

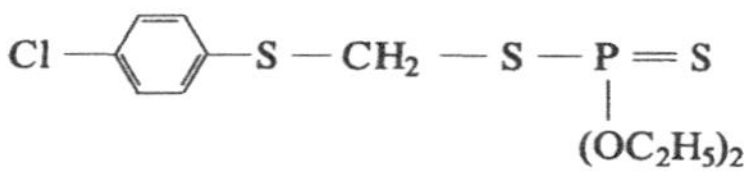

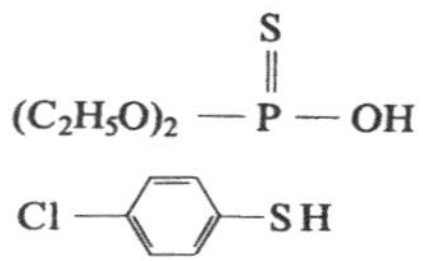

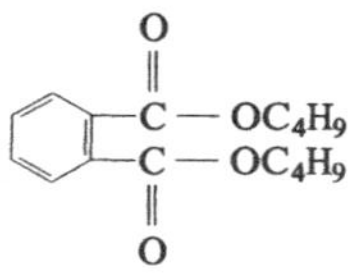

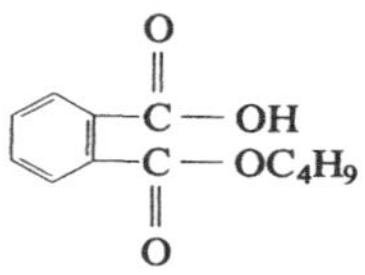

C_6H_4[$C(=O)OC_4H_9$]$_2$ → C_6H_4[$C(=O)OH$][$C(=O)OC_4H_9$] — Foster et al. [30]

b) Bird

Chicken

C_6H_5—$P(=S)$(O—OC_2H_5)(O—C_6H_4—NO_2) → HO—C_6H_4—NO_2 — Lasker et al. [72]

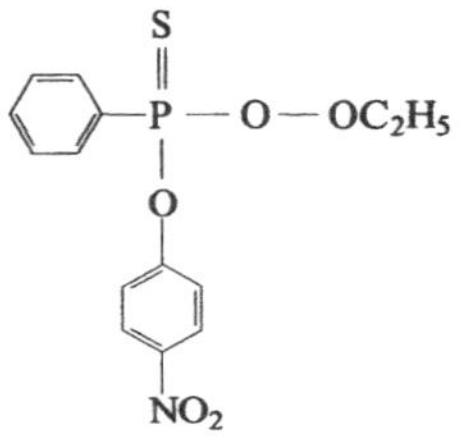

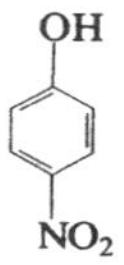

Compound	Metabolite	Reference
c) Fish and Shellfish		
Rainbow trout	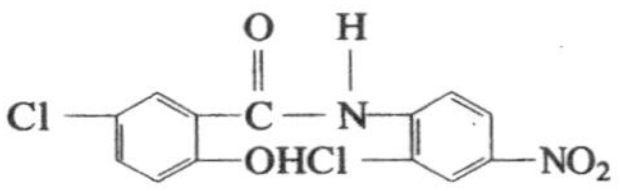 [structure: chlorophenyl amide with NO_2]	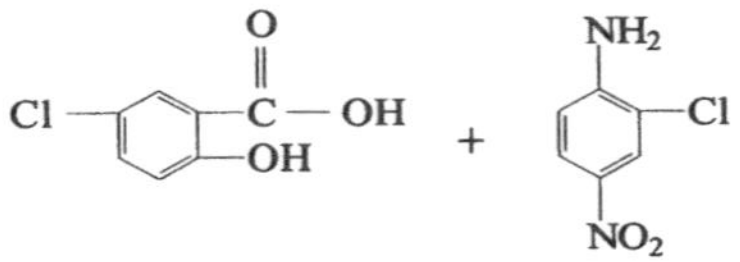 [structure: Cl-benzene-$C(=O)$(OH)(OH) + amino-chloro-nitrobenzene with NH_2, Cl, NO_2]	Allen et al. [5]
Pinfish (*Lagodon rhomboides*)	Malathion	Malathion mono- and di-carboxylic acids	Cook and Moore [19]
Carp, rainbow trout, loach, and shrimp	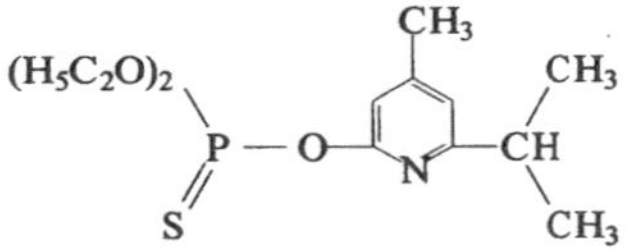 [structure: $(H_5C_2O)_2$P(=S)–O–pyridine with CH_3 groups and $CH(CH_3)_2$]	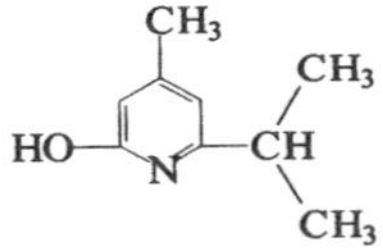 [structure: pyridine ring with HO, CH_3, $CH(CH_3)CH_3$]	Seguchi and Asaka [119]
d) Insect		
Midge larvae (*Chironomus tentans*)	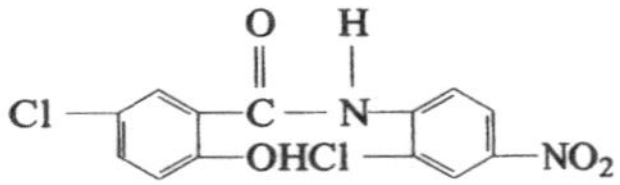 [structure: chlorophenyl amide with NO_2]	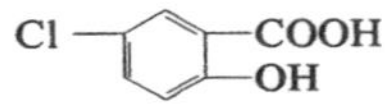 [structure: Cl-benzene-$COOH$, OH]	Kawatski and Zittel [59]
Housefly, flesh fly, and blow fly microsomal enzymes	Hydroprene	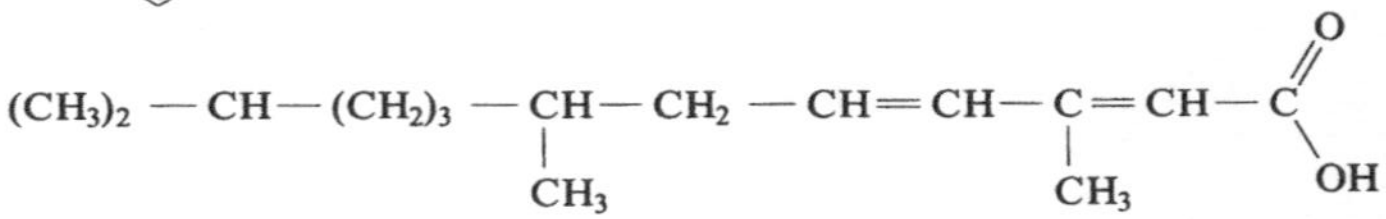 $(CH_3)_2$—CH—$(CH_2)_3$—CH—CH_2—CH=CH—C=CH—C(=O)—OH with CH_3, CH_3	Yu and Terriere [140, 141]
Flour beetle, Colorado beetle, weevil, *Locusta migratoria*, and 8 orders of insects	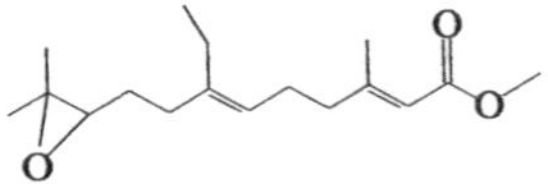 [structure: JH-1]	Free acid and dihydroxy acid	Ajami and Riddiford [4]; Edwards and Rowland [24, 25]; Erley et al. [26]; Kramer et al. [64]; Wilson and Gilbert [138]; Yu and Terriere [142]
e) Plant		
Wheat	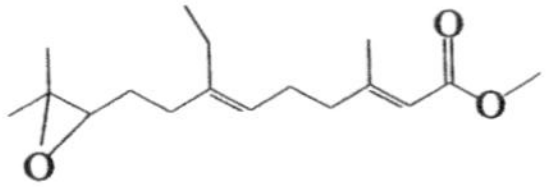 [structure]	Free acid and dihydroxy acid	Rowlands [117]

Organism			Reference
Wild oats	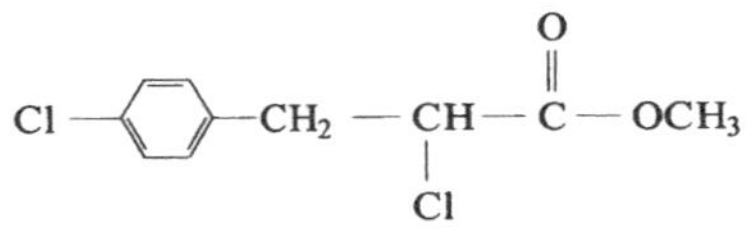 $Cl\text{—}C_6H_4\text{—}CH_2\text{—}CH(Cl)\text{—}C(=O)\text{—}OCH_3$	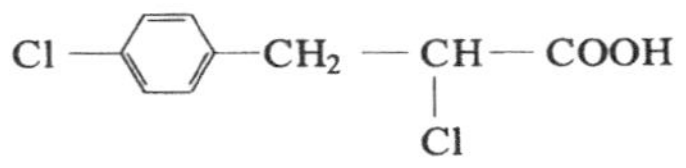 $Cl\text{—}C_6H_4\text{—}CH_2\text{—}CH(Cl)\text{—}COOH$	Fedtke and Schmidt [28]
Rice	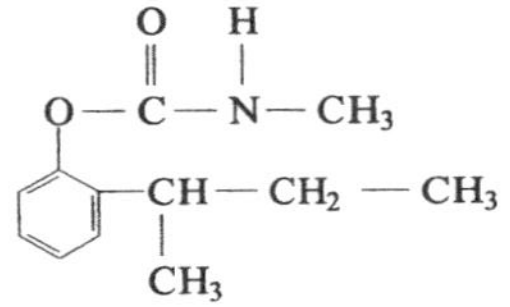 carbamate: $O=C(\text{—}N(H)\text{—}CH_3)\text{—}O\text{—}C_6H_4$ with $CH(\text{—}CH_2\text{—}CH_3)\text{—}CH_3$	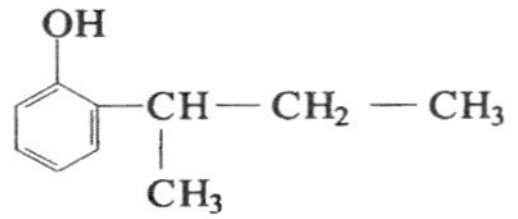 phenol (OH) with $CH\text{—}CH_2\text{—}CH_3$, CH_3	Ogawa et al. [103]

f) Microorganism

Organism			Reference
Flavobacterium sp., *Brevibacterium* sp.	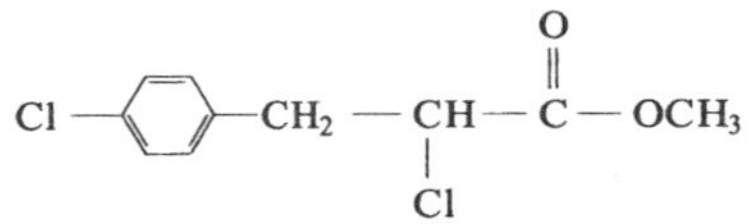 $Cl\text{—}C_6H_4\text{—}CH_2\text{—}CH(Cl)\text{—}C(=O)\text{—}OCH_3$	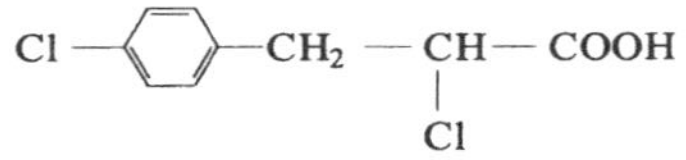 $Cl\text{—}C_6H_4\text{—}CH_2\text{—}CH(Cl)\text{—}COOH$	Kocher et al. [61]
A. Niger, P. Funiculosum, Fusarium sp., *Cladosporium cladosporioides, Coniothyrum* sp.	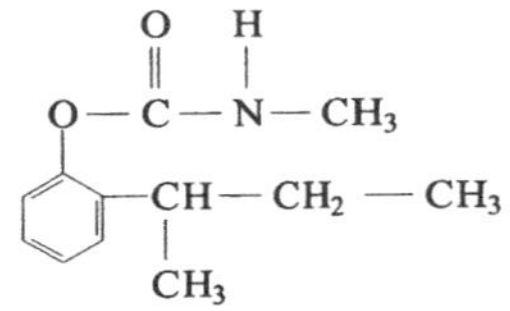 carbamate: $O=C(\text{—}N(H)\text{—}CH_3)\text{—}O\text{—}C_6H_4$ with $CH\text{—}CH_2\text{—}CH_3$, CH_3	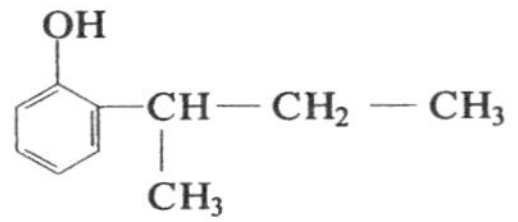 phenol (OH) with $CH\text{—}CH_2\text{—}CH_3$, CH_3	Suzuki and Takeda [126–129]

2. Industrial Products

a) Mammal

Organism			Reference
Rat	Tris(2,3-dibromopropyl)-phosphonate and phosphate	Bis(2,3-dibromopropyl)-phosphonate and phosphate	Lynn et al. [150, 151]

b) Bird

Organism			Reference
Chicks	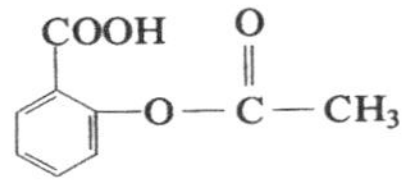 C_6H_4 with COOH and $O\text{—}C(=O)\text{—}CH_3$	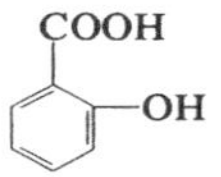 C_6H_4 with COOH and OH	Thomas et al. [152]

Compound	Metabolite	Reference
c) Fish and Shellfish		
Squid nerve enzyme		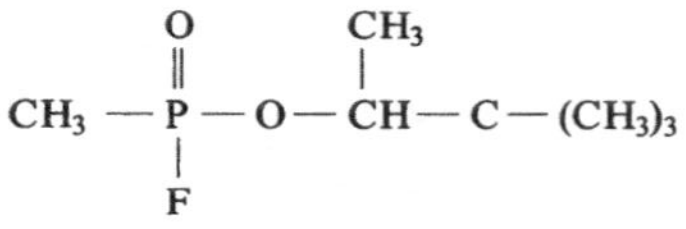 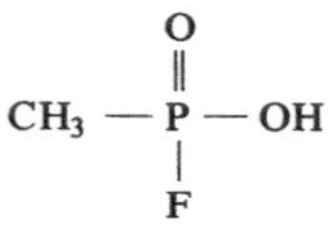 Berry and Davis [11]
Rainbow trout		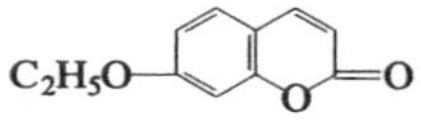 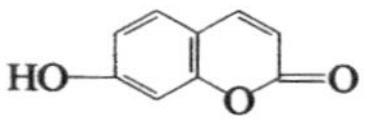 Snegaroff [120]
d) Microorganisms		
Intestinal bacteria Anthraquinone glycosides	Respective anthraquinone analogs	Fingl [153]

H. Dealkylation/Oxidation

1. Agricultural Products

a) Mammal

Rat liver microsomes	&	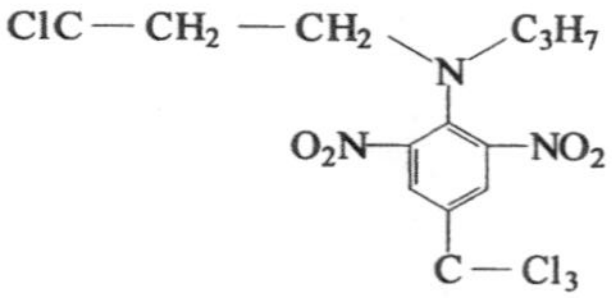 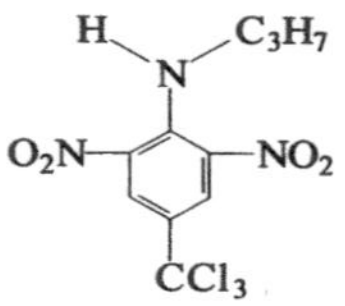 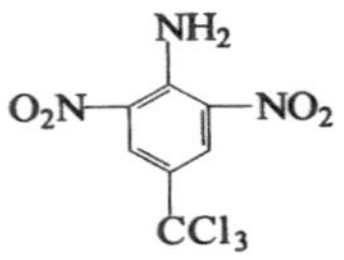 Nelson et al. [97, 98]
Goat	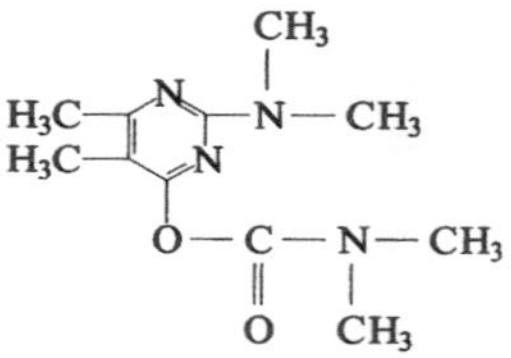 N-Demethyl and N,N-Di-demethyl	Hemingway et al. [41]

b) Bird			
Hens	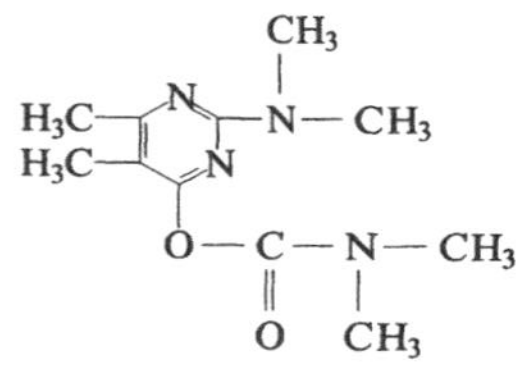	N-Demethyl and N,N-Di-demethyl	Hemingway et al. [41]
c) Fish			
Carp (*Cyprinus carpio*)	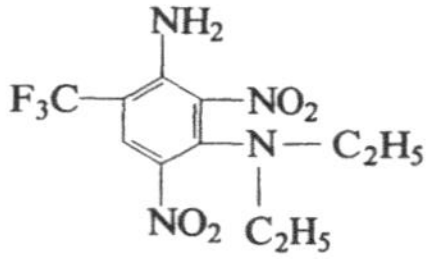	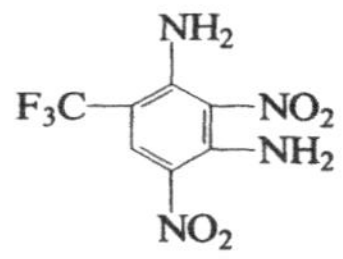	Allen et al. [5]; Olson et al. [106]
S. Mossambicus	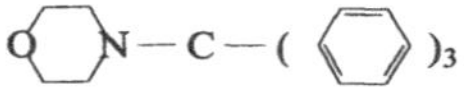	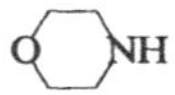	Matthiessen [81]
Rainbow trout	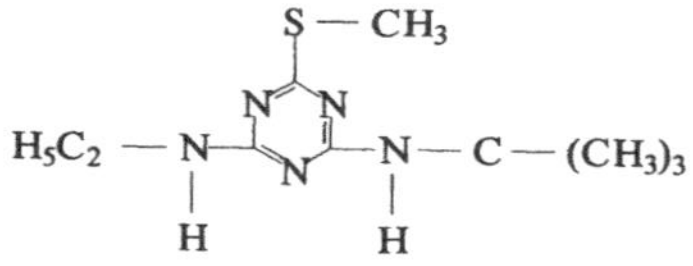	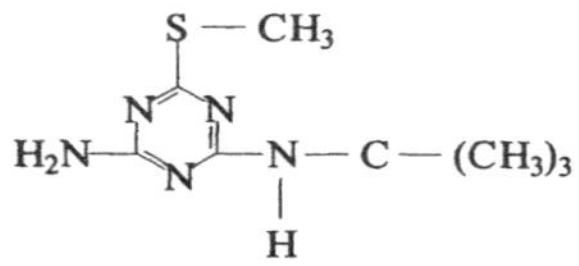	Muir et al. [93]
d) Insect			
Housefly (*Musca domestica*)	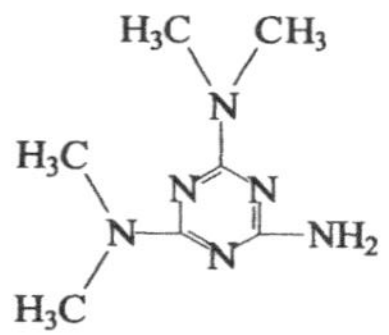	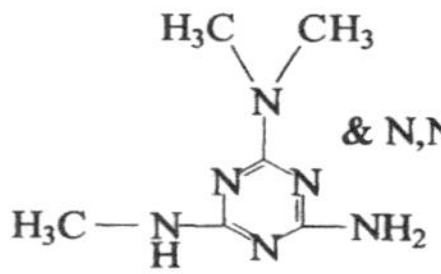 & N,N′-di-demethyl analog	Chang et al. [17]
e) Plant			
Almond seedling	$Cl-CH_2-CH_2-N-(CH_3)_3$	$Cl-CH_2-CH_2-NH_2$	Intrieri and Ryugo [48]

Compound	Metabolite	Reference
f) Microorganism		
Pyricularia oryzae		
$(C_4H_9O)_2 - \overset{\displaystyle O}{\overset{\|}{P}} - N - CH_3$ (phenyl)	$(C_4H_9O)_2 - \overset{\displaystyle O}{\overset{\|}{P}} - NH$ (phenyl)	Uesugi and Sisler [132]
In soils		
$N-(CH_2CH_2Cl)_2$ aromatic ring with O_2N, NO_2, CH_3	NH_2 aromatic ring with O_2N, NO_2, CH_3	Kearney et al. [60]
I. Epoxide Formation		
1. Agricultural Products		
a) Mammal		
Steer	Dieldrin	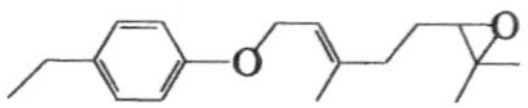 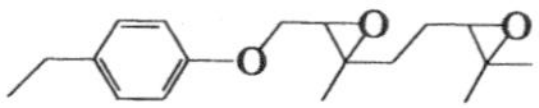 Ivie [49]; Ivie et al. [52]
b) Bird		
Japanese quail	Dieldrin	Gillett and Arscott [37]
c) Fish		
Spot	Heptachlor epoxide	Schimmel et al. [118]
Bass and trout	Dieldrin endrin	Stanton and Khan [123]
d) Insect		
Aldrin	Dieldrin	
Heptachlor	Heptachlor epoxide	
Aldrin, isodrin	Dieldrin endrin	
Housefly, flesh fly, blow fly (*in vivo* and *in vitro*)		
Hydroprene		

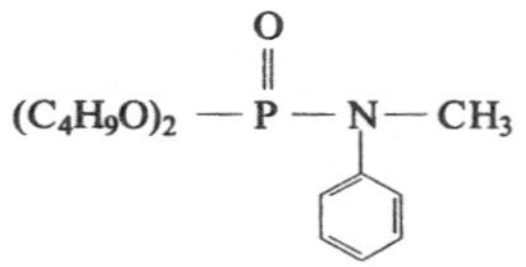

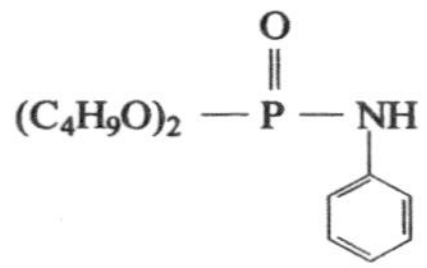

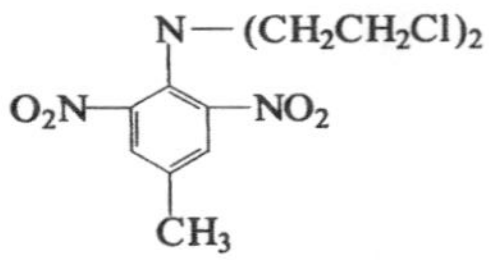

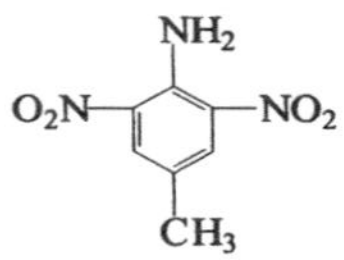

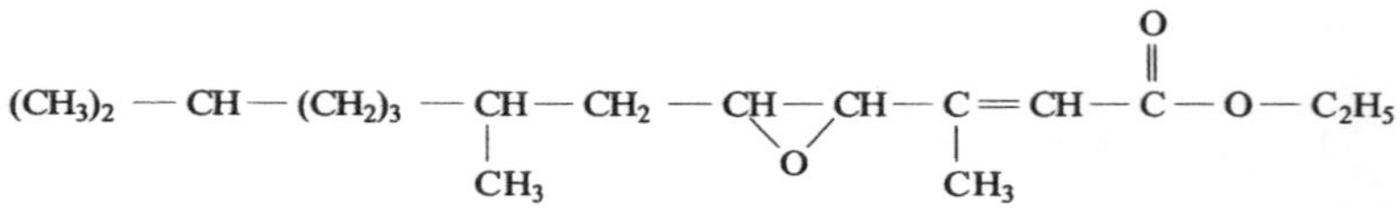

Aldrin structure (ethyl-substituted phenyl ether with methyl-substituted unsaturated chain, terminal gem-dimethyl epoxide)

Dieldrin (metabolite) structure (ethyl-substituted phenyl ether with methyl epoxide and terminal gem-dimethyl epoxide)

Hydroprene:
$(CH_3)_2 - CH - (CH_2)_3 - CH - CH_2 - CH - CH_2 - CH - C = CH - C - O - C_2H_5$ with CH_3, CH_3 substituents, epoxide O, and carbonyl O

Yu and Terriere [140, 141]

Blaberus giganteus

Stable fly

Methyl farnesoate

Methyl farnesoate-10,11-epoxide

Hammock and Mumby [40]

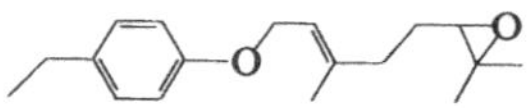

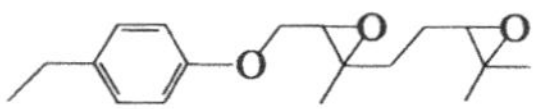

Hammock et al. [39]

2. Industrial Products

a) Mammal

Mouse hepatic microsomes

Gervasi et al. [36]

Rat

$CH_2 = CH - CHO$

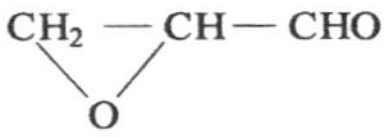

$CH_2 - CH - CHO$

Van and Loewengart [154]

J. Exchange Reaction

1. Agricultural Products

a) Mammal

Dog

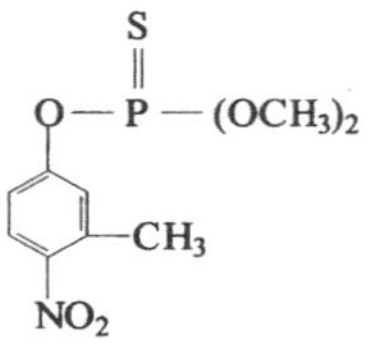

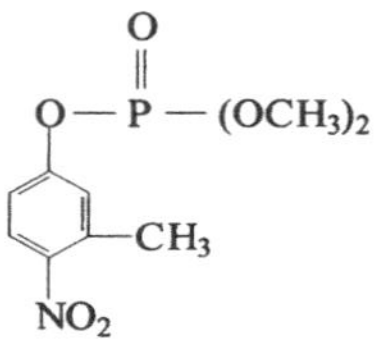

$S = P - (OCH_3)_2$

$O = P - (OCH_3)_2$

Miyamota et al. [90]

b) Bird

Chicken

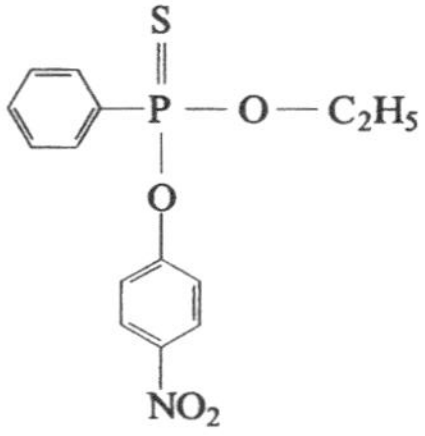

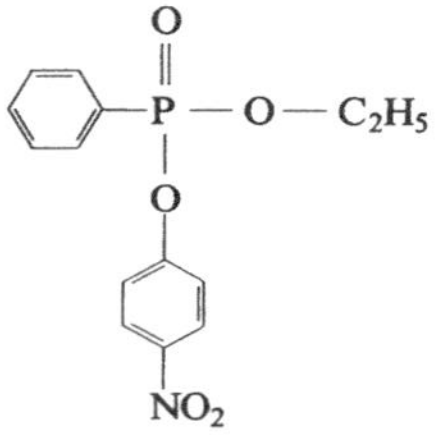

$S = P - O - C_2H_5$

$O = P - O - C_2H_5$

Lasker et al. [72]

Compound	Metabolite	Reference
c) Fish		
Channel catfish liver microsomes $(H_5C_2O)_2$ (structure)	H_5C_2O, $O=P$, H_5C_2O (structure)	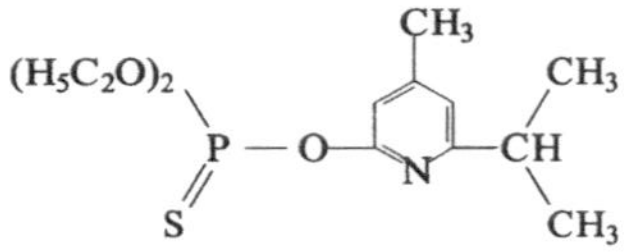 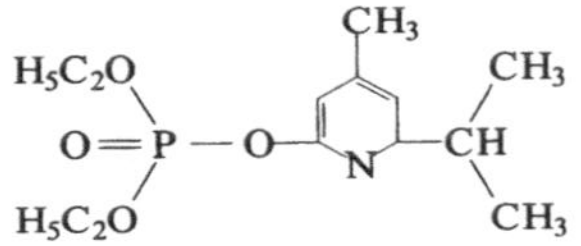 Hogan [43]
d) Insect		
Egyptian cotton leafworm larvae (structure)	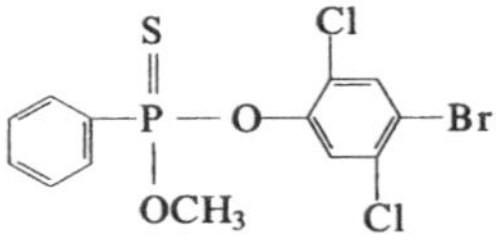 (structure)	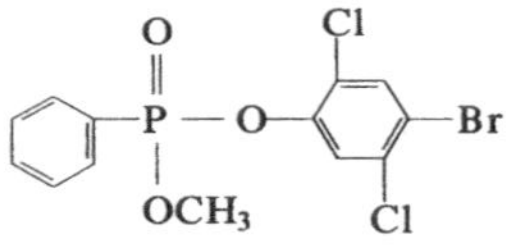 Riskallah and Esaac [114]

K. Photolytic Effects

Photochemical reaction of nitrogen oxides with fluorocarbons, which find their way into the air, may result in the formation of hydrogen fluoride and hydrogen chloride; and when vinyl chloride and ethyl chloride were subjected to photolysis in the presence of nitrogen oxides in air, formaldehyde and hydrogen chloride were produced [56, 57]. When a neutral solution of nitrate and biphenyl was irradiated as a dinitrodihydroxybiphenyl [125]. Under natural environmental conditions, dimethylamine was nitrosated by exposure to sunlight in an aqueous solution containing nitrite. The product, nitrosodimethylamine, is a wellknown carcinogen [155]. The heterocyclic acridine was found to be harmless to paramecia in the absence of light. However, when the organisms were exposed to acridine in the presence of visible light, then acridine was lethal [111]. In other studies, seeds of the perenial desert plant, *Pituranthos triradiatus* (Hochst. ex Boiss.) Aschers et Schweinf, occasionally grazed upon by goats, gazelles and camels, was fed to ducklings. When the birds was exposed to sunlight, severe photosensitization occurred and pathological lesions observed were similar to those seen in photosensitized ducklings in other studies. Examination of the seeds revealed the presence of furocoumarins, used in treating psoriasis in man with UV [7].

1. Agricultural Products

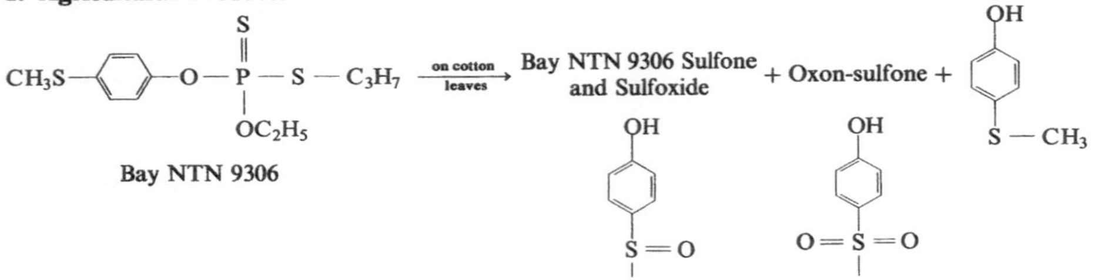

CH_3S — (structure) — O—P—S—C_3H_7 $\xrightarrow[\text{leaves}]{\text{on cotton}}$ Bay NTN 9306 Sulfone + Oxon-sulfone + Sulfoxide and Sulfoxide

Bay NTN 9306

(structures: OH ... S—CH_3; OH ... $O=S=O$; OH ... $S=O$)

Ivie and Bull [50]

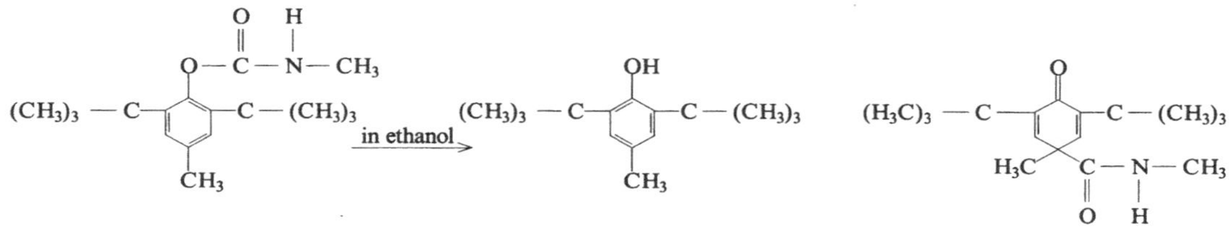

O H
O—C—N—CH3
(CH3)3—C
C—(CH3)3
CH3
in ethanol
OH
(CH3)3—C
C—(CH3)3
CH3
O
(H3C)3—C
C—(CH3)3
H3C
C—N—CH3
O H

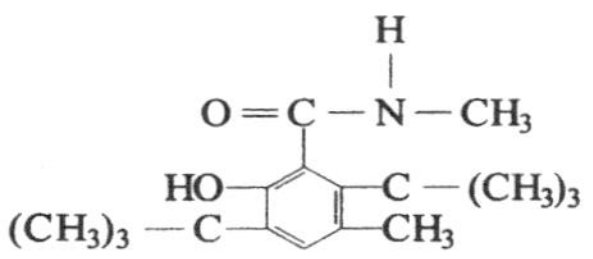

H
O=C—N—CH3
HO
C—(CH3)3
(CH3)3—C
CH3

Kumar et al. [67]

2. Industrial Products

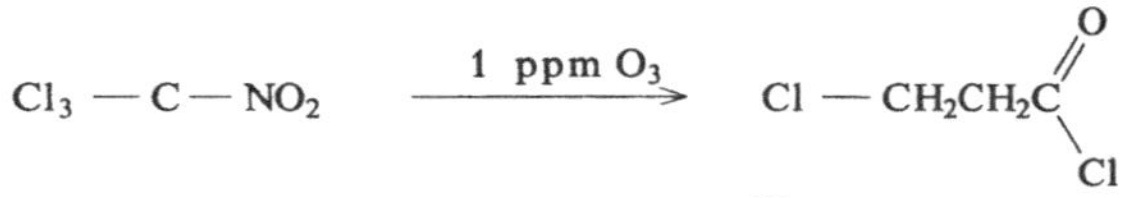

Cl3—C—NO2
1 ppm O3
Cl—CH2CH2C
O
Cl

Moilanen et al. [92]

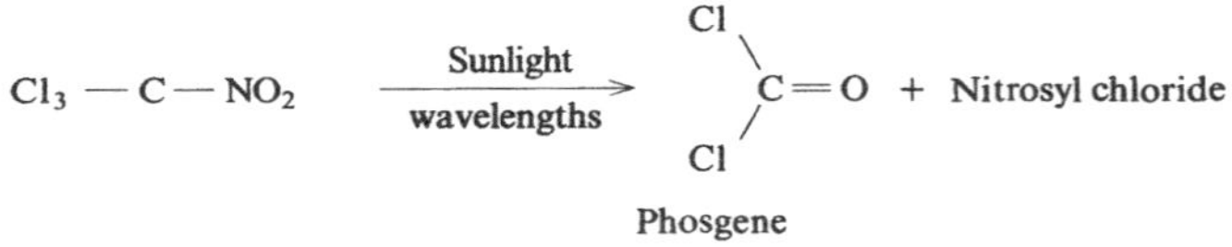

Cl3—C—NO2
Sunlight
wavelengths
Cl
C=O + Nitrosyl chloride
Cl
Phosgene

Moilanen et al. [91]

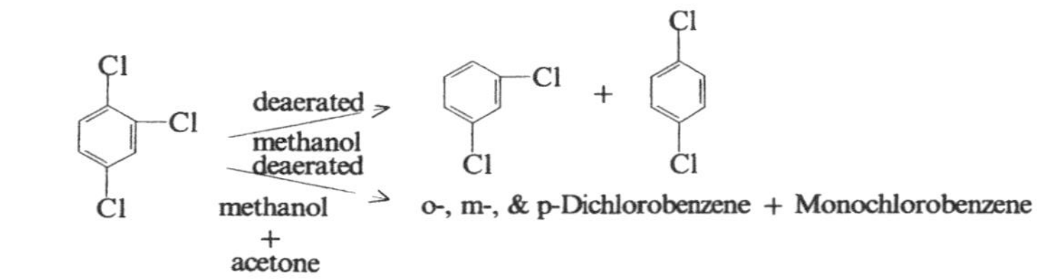

Cl
Cl
Cl
deaerated
methanol
deaerated
methanol
+
acetone
Cl
Cl
+
Cl
Cl
o-, m-, & p-Dichlorobenzene + Monochlorobenzene

Choudry et al. [18]

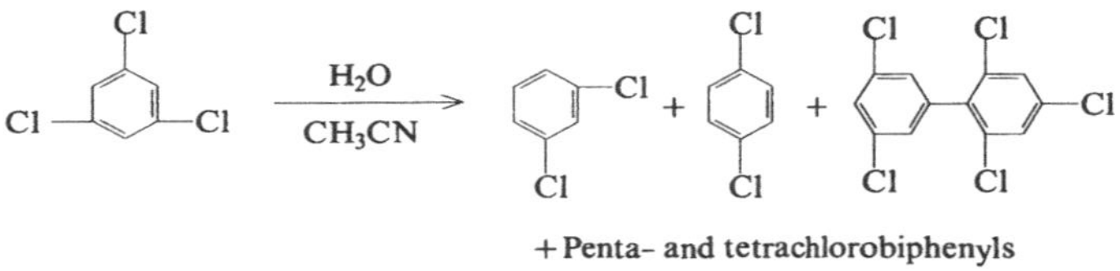

Cl
Cl
Cl
H2O
CH3CN
Cl
Cl
+
Cl
Cl
+
Cl Cl
Cl
Cl Cl
+Penta- and tetrachlorobiphenyls

Choudry et al. [18]

References

1. Adamson, R.H., et al.: Comparative biochemistry of drug metabolism by azo and nitro reductase. Proc. Natl. Acad. Sci. U.S.A. *54*, 1386 (1965)
2. Ahlborg, U.G.: Dechlorination of pentachlorophenol *in vivo* and *in vitro*. *in* Pentachlorophenol, 115–130. Ed. K. Ranga Rao. Published by Plenum Publ. Corp. 1978
3. Ahlborg, U.G., Larsson, K., Thunberg, T. Metabolism of pentachlorophenol *in vivo* and *in vitro*. Arch. Toxicol. *40*, 459 (1978)
4. Ajami, A.M., Riddiford, L.M.: Comparative metabolism of the Cecropia juvenile hormone. J. Insect Physiol. *19*, 635 (1973)
5. Allen, J.L., Hunn, J.B., Dawson, V.K.: Biotransformation of selected chemicals by fish. Abstr. 176th ACS Meeting, Miami Beach, FL, PEST 72 (1978)
6. Aschbacher, P.W., Feil, V.J.: Mechanism of pentachloronitrobenzene in the goat. Abstr. 172nd ACS Meeting, San Francisco, CA, PEST 92 (1976)
7. Ashkenazy, D., et al.: Photosensitization in ducklings induced by Pituranthos Triradiatus. Vet. Human Toxicol. *26* (2), 118 (1984)
8. Avrahami, M., White, D.A.: Rapid elimination of pentachloronitrobenzene and metabolite from sheep fat. N.Z.J. Exper. Agric. *4*, 299 (1976)
9. Baldwin, M.K., et al.: The metabolism and residues of [^{14}C]endrin in lactating cows and laying hens. Pestic. Sci. *7*, 575 (1976)
10. Berninger, V.H., Ammon, R., Berninger, I.: Spektrophotometrische Identifizierungen von Hydroxyderivaten des Biphenyls. Arzneim. Forsch. *18*, 880 (1968)
11. Berry, R.W., Davis, D.R.: Factors influencing the rate of aging of a series of alkyl methylphosphonylacetylcholinesterase. Biochem. J. *100*, 572 (1966)
12. Bridges, J.W., et al.: The conjugation of indolylacetic acid in man, monkeys and other species. Xenobiotica *4* (10), 645 (1974)
13. Bull, D.L.: Fate and efficacy of sulprofos against certain insects associated with cotton. J. Econ. Entomol. *73*, 262 (1980)
14. Bull, D.L., Ivie, G.W.: Metabolism of O-ethyl O-[4-(methylthio)phenyl] S-propyl phosphorodithioate (BAY NTN 9306) by white rats. J. Agric. Food Chem. *24* (1), 143 (1976)
15. Cannizzaro, R.D.: Residues of bentazon and its 6- und 8-hydroxylated metabolites in soybeans and soil. Abstr. Weed Sci. Soc. Amer. 82 (1975)
16. Capel, J.D., Millburn, P., Williams, R.T.: Route of administration and the conjugation of phenol in hens. Biochem. Soc. Trans. *2*, 875 (1974)
17. Chang, S.C., Woods, C.W., Borkovec, A.B.: Metabolism of ^{14}C-labeled N^2,N^2,N^4,N^4-tetramethylmelamine in male house flies. J. Econ. Entomol. *63* (5), 1510 (1970)
18. Choudhry, G.G., Roof, A.A.M., Hutzinger, O.: Photochemistry of halogenated benzene derivatives. I. Trichlorobenzenes: Reductive dechlorination, isomerization and formation of polychlorobiphenyls. Tetrahedron Lett. No. 22, 2059 (1979)
20. DeBaun, J.R., Menn, J.: Sulfoxide reduction in relation to organophosphorus insectizide detoxification. Sci. *191*, 187 (1976)
21. Douch, P.G.C., Gahagan, H.M.: The metabolism of niclosamide and related compounds by *Moniezia expansa, Ascaris lumbricoides* var. *suum*, and mouse and sheep liver enzymes. Xenobiotica *7* (5), 301 (1977)
22. Drager, G.: Studies on the isolation and identification of extractable and nonextractable metabolites of the insecticide carbamate croneton. Abstr. 4th Internat. Congr. Pest. Chem. (IUPAC), Zurich, V-23 (1978)
23. Dunn, J.S., et al.: The accumulation and elimination of tissue residues after feeding pentachloronitrobenzene to white leghorn cockerels. Poultry Sci. *57* (6), 1533 (1978)
24. Edwards, J.P., Rowlands, D.G.: Metabolism of a synthetic juvenile hormone (JH-1) during the development of *Tribolium castaneum* (Herbst) (Coleoptera, Tenebrionidae). Pestic. Biochem. Physiol. *7*, 194 (1977)
25. Edwards, J.P., Rowlands, D.G.: Metabolism of a synthetic juvenile hormone (JH-1) in two strains of the grain weevil, *Sitophilus granarius*. Insect Biochem. *8*, 23 (1978)
26. Erley, D., Southard, S., Emmerich, H.: Excretion of juvenile hormone and its metabolites in locust, *Locusta migratoria*. J. Insect Physiol. *21*, 61 (1975)

27. Farooqui, M.Y.H., Metcalf, R.L.: Comparative metabolism of six ethoxychlor analogs and DDT in DDT-resistant and susceptible strains of the house fly *Musca domestica* L. Pestic. Biochem. Physiol. *19*, 210 (1983)
28. Fedtke, C., Schmidt, R.R.: Chlorfenprop-methyl: Its hydrolysis *in vivo* and *in vitro* and a new principle for selective herbicidal action. Weed Res. *17*, 233 (1977)
29. Fletcher, C.L., Kaufman, D.D.: Hydroxylation of monochloroaniline pesticide residues. Abstr. Weed. Sci. Soc. Amer. *76* (1975)
30. Foster, R.M.D., et al.: Differences in urinary metabolic profile from di-n-butyl phthalate-treated rats and hamsters. Drug Metab. Dispos. *11*, 59 (1983)
31. Foster, G.D., Crosby, D.G.: The xenobiotic metabolism of methyl parathion by decapod crustaceans. Abstr. 188th ACS Meeting, Phila., Penn., PEST 17 (1984)
32. Fouts, J.R., Kamm, J.J., Brodie, B.B.: Enzymatic reduction of prontosil and other azo dyes. J. Pharmacol. Expt. Ther. *120*, 291 (1957)
33. Frear, D.S., Swanson, H.R.: Metabolism of cisanilide (cis-2,5-dimethyl-1-pyrrolidine-carboxanilide) by excised leaves and cell suspension cultures of carrot and cotton. Pestic. Biochem. Physiol. *5*, 73 (1975)
34. Frear, D.S., Swanson, H.R., Davison, K.L.: Cisanilide (cis-2,5-dimethyl-1-pyrrolidine-carboxanilide) metabolism in plants: Isolation, characterization and degradation of insoluble residue fraction. 170th ACS Meeting, Chicago, IL, PEST 11 (1975)
35. Gardner, A.M., Alvarez, G.H., Ku, Y.: Microbial degradation of ^{14}C-diphenylamine in a laboratory model sewage sludge system. Bull. Environm. Contam. Toxicol. *28*, 91 (1982)
36. Gervasi, P.G., et al.: The metabolism of 1,3-cyclohexadiene by liver microsomal mono-oxygenase. Xenobiotica *12* (8), 517 (1982)
37. Gillett, J.W., Arscott. G.H.: Microsomal epoxidation in Japanese quail: Induction by dietary dieldrin. Comp. Biochem. Physiol. *30*, 589 (1969)
38. Hajjar, N.P., et al.: S-(O-Ethyl phenylphosphonothionyl)glutathione; evidence for its formation in the *in vitro* metabolism of EPN in houseflies. Drug and Chem Toxicol. *3* (4), 421 (1980)
39. Hammock, B.D., et al.: Metabolic O-dealkylation of 1-(4-ethylphenoxy)-3,7-dimethyl-7-methoxy or ethoxy-trans-2-octene, potent juvenoids. Pestic. Biochem. Physiol. *5*, 12 (1975)
40. Hammock, B.D., Mumby, S.M.: Inhibition of epoxidation of methyl farnesoate to juvenile hormone III by cockroach corpus allatum homogenates. Pestic. Biochem. Physiol. *9*, 39 (1978)
41. Hemingway, R.J., et al.: Residue transfer and metabolism of pirimicarb in hens and a goat. Abstr. 4th Internat. Congr. Pest. Chem, (IUPAC), Zurich, V-518 (1978)
42. Hicks, B.W., Dorough, H.W., Mehendale, H.M.: Metabolism of aldicarb pesticide in laying hens. J. Agric. Food Chem. *20*, 151 (1972)
43. Hogan, J.W., Knowles, C.O.: Metabolism of diazinon by fish liver microsomes. Bull. Environm. Contam. Toxicol. *8*, 61 (1972)
44. Hoodi, A.A., Damani, L.A.: Sulphoxidation of alicyclic and aliphatic sulphides. First Internat. Sympos. Foreign Cmpd. Metab. Abstr., p. 34 (1983)
45. Hunt, L.M., et al.: Absorption, excretion, and metabolism of nitrofen by a sheep. J. Agric. Food Chem, *25* (5), 1062 (1977)
46. Hutacharern, C.: Action and metabolism of chlorpyrifos in termites. Dissert. Abstr. 36B (3), 1057 (1975)
47. Hutacharern, C., Knowles, C.O.: Metabolism of chlorpyrofos-^{14}C in the eastern termite. Bull. Environm. Contam. Toxicol. *13* (3), 351 (1975)
48. Intrieri, C., Ryugo, K.: Uptake, transport, and metabolism of (2-chloroethyl)trimethyl-ammonium chloride in seedlings of almond (*Prunus amygdalus*, Batsch). J. Amer. Soc. Hort. Sci. *99* (4), 349 (1974)
49. Ivie, G.W.: Epoxide to olefin: A novel biotransformation in the rumen. Sci. *191*, 959 (1976)
50. Ivie, G.W., Bull, D.L.: Photodegradation of O-ethyl O-[4-(methylthio)phenyl] S-propyl phosphorodithioate (Bay NTN 9306). J. Agric. Food Chem, *24* (5), 1053 (1976)
51. Ivie, G.W., Bull, D.L., Witzel, D.A.: Metabolite fate of O-ethyl O-[4-(methylthio)phenyl-^{14}C] S-propyl phosphorodithioate (Bay NTN 9306) in a lactating cow. J. Agric. Food Chem. *24* (1), 147 (1976)

52. Ivie, G.W., Wright, J.E., Smalley, H.E.: Fate of the juvenile hormone mimic 1-(4'-ethylphenoxy)-3,7-dimethyl-6,7-epoxy-*trans*-2-octene (Stauffer R-20458) following oral and dermal exposure to steers. J. Agric. Food Chem. *24* (2), 222 (1976)

53. James, M.O.: Disposition and taurine conjugation of 2,4-dichlorophenoxyacetic acid, 2,4,5-trichlorophenoxyacetic acid, bis(4-chlorophenyl)acetic acid, and phenylacetic acid in the spiny lobster, *Panulirus argus*. Drug Metab. Dispos. *10* (5), 516 (1982)

54. Kadous, A.A.R.: Chlorinated hydrocarbon insecticides: Mode of action, resistance mechanisms, and metabolism. Dissert. Abstr. 39B (6), 2652 (1978)

55. Kanazawa, J., Tomozawa, C.: Intake and excretion of 2,4,6-trichlorophenyl-4'-nitrophenyl ether by topmouth gudgeon, *Pseudorasboro parva*. Arch. Environm. Contam. Toxicol. *7*, 397 (1978)

56. Kanno, S., Ito, T., Omura, T.: Studies on the photochemistry of aliphatic halogenated hydrocarbons. Chemosphere *6* (8), 503 (1977)

57. Kanno, S., et al.: Studies on the photochemistry of aliphatic halogenated hydrocarbons II. Chemosphere *6* (8), 509 (1977)

58. Karanth, N.G.K.: Degradation of Dexon (p-dimethylaminobenzenediazo sodium sulfonate to N,N-dimethyl-p-phenylenediamine by *Pseudomonas fragi* BK₉. Appl. Microbiol. *27*, 43 (1974)

59. Kawatski, J.A., Zittel, A.E.: Accumulation, elimination, and biotransformation of the lampricide 2',5-dichloro-4'-nitrosalicylanilide by *Chironomus tentans*. U.S. Fish Wildl. Serv. Invest. Fish Control *79* (1977)

60. Kearney, P.C., et al.: Persistence and metabolism of dinitroaniline herbicides in soils. Pestic. Biochem. Physiol. *6*, 229 (1976)

61. Kocher, H., Lingens, F., Koch, W.: Untersuchungen zum Abbau des Herbizids Chlorophenpropmethyl im Boden und durch Mikroorganismen. Weed Res. *16*, 93 (1976)

62. Koizumi, A., Kumai, M., Ikeda, M.: Enzymatic formation of an olefin in the metabolism of 1,1,2,2-tetrachloroethane: an *in vitro* study. Bull. Environm. Contam. Toxicol. *29*, 562 (1982)

63. Krahn, M.M., et al.: Determination of metabolites of xenobiotics in the bile of fish from polluted waterways. Xenobiotica *14* (8), 633 (1984)

64. Kramer, S.J., Wieten, M., de Kort, C.A.D.: Metabolism of juvenile hormone in the Colorado potato beetle, *Leptinotarsa decemlineata*. Insect Biochem. *7*, 231 (1977)

65. Krause, M., Van Dyk, L.P.: Persistence of phorate and its sulfoxide in cabbages. Phytophylactica *7*, 119 (1975)

66. Krieger, R.I., et al.: Aldrin epoxidation, dihydroisodrin hydroxylation and p-chloro-N-methylaniline demethylation in six species of *Saturniid* larvae. J. Econ. Entomol. *69* (1), 1 (1976)

67. Kumar, Y., et al.: The photochemistry of carbamates. IV. The phototransformation of azak (2,6-di-t-butyl-4-methylphenyl-N-methylcarbamate. Environ. Qual. Safety 3 (Suppl.), 302 (1974)

68. Kupelian, R.H.: Bentazon metabolism in soybeans, animals, and soil. Abstr. Weed Sci. Soc. Amer. 82 (1977)

69. Kuwatsuka, S., Igarashi, M.: Degradation of PCP in soils. II. The relationship between the degradation of PCP and the properties of soils, and the identification of the degradation products of PCP. Soil Sci. Plant Nutr. *21* (4), 405 (1975)

70. Kuwatsuka, S., Oyamada, M., Shimotori, H.: Metabolic fate of CNP, a diphenyl ether herbicide, in soil and by isolated soil microorganisms. Abstr. 4th Internat. Congr. Pest. Chem. (IUPAC), Zurich, V-504 (1978)

71. Lamoureux, G.L., Rusness, D.G.: Pentachloronitrobenzene (PCNB) metabolism in peanut. Abstr. 172nd ACS Meeting, San Francisco, CA, PEST 106 (1976)

72. Lasker, J.M., Graham, D.G., Abou-Donia, M.B.: Differential metabolism of O-Ethyl-O-4-nitrophenyl phenylphosphonothioate by rat and chicken hepatic microsomes. Biochem. Pharmacol. *31* (11), 1961 (1982)

73. LePatourel, G.N.J., Wright, D.J.: Organophosphorus pesticide metabolism associated with nematodes. The role of bacteria. Nematologica *20*, 104 (1974)

74. Lichtenstein, E.P., Liang, T.T., Fuhreann, T.W.: A compartmentalized microcosm for studying the fate of chemicals in the environment. J. Agric. Food Chem. *26* (4), 948 (1978)

75. Lores, E.M., et al.: A new metabolite of chlorpyrifos: Isolation and identification. Abstr. 173rd ACS Meeting, New Orleans, LA, PEST 50 (1977)

76. Lores, E.M., et al.: A new metabolite of chlorpyrifos: Isolation and identification. J. Agric. Food Chem. 26 (1), 118 (1978)

77. Luke, B.G., Dahl, C.J.: Detection of O,O-diethyl-O-(2,5-dichlorophenyl)phosphorothioate and O,O-diethyl-O-(3,6-dichloro-2-pyridyl)phosphorothioate in meat fat. J. Assoc. Off. Anal. Chem. 59 (5), 1081 (1976)

78. Manis, M.O., Braselton, W.E.: Metabolism of 4,4'-methylenebis(2-chloroaniline) by dog tissue slices. Abstracts First Internat. Sympos. Foreign Cmpd. Metabol. p. 44 (1983)

79. Marshall, T.C., Dorough, H.W.: Bioavailability in rats of bound and conjugated plant carbamate insecticide residues. J. Agric. Food Chem. 25 (5), 1003 (1977)

80. Mate, C., Ryan, A.J., Wright, S.E.: Metabolism of some 4-nitroaniline derivatives in the rat. Food Cosmet. Toxicol. 5, 657 (1967)

81. Matthiessen, P.: Uptake, Metabolism, and excretion of the molluscide N-tritylmorpholine by the tropical food fish Sarotherodon mossambicus (Peters). J. Fish Biol. 11, 487 (1977)

82. Menn, J.J., et al.: Metabolism of thioaryl organophosphorus insecticides. Environ. Qual. Safety 3 (Suppl.), 382 (1974)

83. Meshi, T., Yoshikawa, M., Sato, Y.: Metabolic fate of bis(3,5-dichloro-2-hydroxyphenyl)-sulfoxide (bithional sulfoxide). Biochem. Pharmacol. 19, 1070 (1970)

84. Metcalf, R.L., et al.: Model ecosystem studies of the environmental fate of six organo-chlorine pesticides. Environ. Hlth. Perspec. 4, 35 (1973)

85. Meyer, T.: The metabolism of biphenyl. IV. Phenolic metabolites in the guinea pig and the rabbit. Acta Pharmacol. Toxicol. 40, 193 (1977)

86. Meyer, T., Aarbakke, J., Scheline, R.R.: The metabolism of biphenyl. I. Metabolic disposition of ^{14}C-biphenyl in the rat. Acta Pharmacol. Toxicol. 39, 412 (1976)

87. Meyer, T., Scheline, R.R.: The metabolism of biphenyl. II. Phenolic metabolites in the rat. Acta Pharmacol. Toxicol. 39, 419 (1976)

88. Mihara, K., Misaki, Y., Miyamoto, J.: Metabolism of Fenitrothion in birds. J. Pestic. Sci. 4, 175 (1979)

89. Mine, A., Miyakado, M., Matsunaka, S.: The mechanism of bentazon selectivity. Pestic. Biochem. Physiol. 5, 566 (1975)

90. Miyamoto, J., Mihara, K., Hosokawa, S.: Comparative metabolism of m-methyl-^{14}C-sumithion in several species of mammals in vivo. J. Pestic. Sci 1 (1), 9 (1976)

91. Moilanen, K.W., et al.: Vapor-phase decomposition of fumigants. Abstr. 172nd ACS Meeting, San Francisco, CA, PEST 98 (1976)

92. Moilanen, K.W., et al.: Vapor-phase photodecomposition of pesticides in the presence of oxidant. Abstr. 173rd ACS Meeting, New Orleans, LA, PEST 2 (1977)

93. Muir, D.C.G., et al.: Comparison of the uptake and bioconcentration of Fluridone and Terbutryn by rainbow trout and Chironomus tentans in sediment and water systems. Arch. Environm. Contam. Toxicol. 11, 595 (1982)

94. Muller, W.F., et al.: Comparative metabolism of hexachlorobenzene and pentachloronitro-benzene in plants, rats, and rhesus monkeys. Ecotoxicol. Environ. Safety 1, 437 (1978)

95. Nakagawa, M., Ando, M.: Metabolism of credazine, 3-(2'-methylphenoxy)pyridazine, in rats. Agric. Biol. Chem. 41 (10), 1975 (1977)

96. Nakagawa, M., Kawakuko, Ishida, M.: Metabolism of the herbicide 3-(2'-methylphenoxy)-pyridazine in plants. Agric. Biol. Chem. 35 (5), 764 (1971)

97. Nelson, J.O., et al.: Metabolism of three dinitroaniline herbicides by rat liver microsomes. Abstr. 172nd ACS Meeting, San Francisco, CA, PEST 59 (1976)

98. Nelson, J.O., et al.: Metabolism of trifluralin, profluralin, and fluchloralin by rat liver microsomes. Pestic. Biochem. Physiol. 7, 73 (1977)

99. Niki, Y., Kuwatsuka, S.: Degradation of diphenyl ether herbicides in soils. Soil Sci. Plant Nutr. 22 (3), 223 (1976)

100. Niki, Y., Kuwatsuka, S.: Degradation products of chlomethoxynil (X-52) in soil. Soil Sci. Plant Nutr. 22 (3), 233 (1976)

101. Niki, Y., Kuwatsuka, S., Yokomichi, I.: Absorption, translocation and metabolism of chlomethoxynil (X-52) in plants. Agric. Biol. Chem. 40 (4), 683 (1976)

102. Nye, D.E., Hurst, H.E., Dorough, H.W.: Fate of Croneton (2-ethylthiomethylphenyl N-methylcarbamate) in rats. J. Agric. Food Chem. *24* (2), 371 (1976)
103. Ogawa, K., et al.: Metabolism of 2-sec-butylphenyl N-methylcarbamate (Bassa, BPMC) in rice plants and its degradation in soils. J. Pestic. Sci. *1*, 219 (1976)
104. Ohkawa, H., Okihara, Y., Miyamoto, J.: Metabolism of the fungicide Denmert (S-n-butyl S'-p-tert-butylbenzyl N-3-pyridyldithiocarbonimidate, S-1358) and Denmert sulfoxides in liver enzyme systems. Agric. Biol. Chem. *40* (6), 1175 (1976)
105. Ohta, T., et al.: Photochemical nitrosation of dimethylamine in aqueous solution containing nitrite. Chemosphere *11* (8), 797 (1982)
106. Olson, L.E., Allen, J.L., Hogan, J.W.: Biotransformation and elimination of the herbicide dinitramine in carp. J. Agric. Food Chem. *25* (3), 554 (1977)
107. Otto, S., et al.: Investigations into the degradation of bentazon in plant and soil. 4th Internat. Congr. Pest. Chem., Zurich in *Advances in Pesticide Science,* ed. H. Geiss-buhler, Pergamon Press, Part 3, p. 551 (1978)
108. Pan, H.P., Hook, G.E.R., Fouts, J.R.: The liver parenchyma and foreign compound metabolism in red-winged blackbird compared with rat. Xenobiotica *5* (1), 17 (1975)
109. Patterson, D.S.P., Roberts, B.A.: The in vitro reduction of aflatoxins B_1 and B_2 by soluble liver enzymes. Food Cosmet. Toxicol. *9*, 829 (1971)
110. Poiger, H., et al.: Structure elucidation of mammalian TCDD metabolites. Experientia *38*, 484 (1982)
111. Raab, O.: Ueber die Wirkung fluorescirender Stoffe auf Infusorien. Z. Biol. *39*, 524 (1900)
112. Reed, D.L., et al.: Tissue residues from feeding pentachloronitrobenzene to broiler chickens. Toxicol. Appl. Pharmacol. *42*, 433 (1977)
113. Renner, G., et al.: Isolierung von Stoffwechselprodukten der Pestizide Hexachlorbenzol und Pentachlornitrobenzol aus Harn und Faeces von Ratten und Kaninchen. Chemosphere *7* (12), 943 (1978)
114. Riskallah, M.R., Esaac, E.G.: Metabolism and excretion of Leptophos and its phenol in S and R larvae of the Egyptian cotton leafworm. Arch. Environm. Contam. Toxicol. *11*, 253 (1982)
115. Riov, J., Cooper, R., Gottlieb, H.E.: Metabolism of auxin in pine tissues: Naphthalene-acetic acid conjugation. Physiol. Plant. *46*, 133 (1979)
116. Robocker, W.C., Zamora, B.A.: Translocation and metabolism of dicamba in western bracken. Weed Sci. *24* (4), 435 (1976)
117. Rowlands D.G.: The uptake and metabolism by stored wheatgrains of an insect juvenile hormone and two insect hormone mimics. J. Stored Prod. Res. *12*, 35 (1976)
118. Schimmel, S.C., Patrick, J.M., Forester, J.: Heptachlor: Uptake, depuration, retention and metabolism by spot, *Leiostomus xanthurus.* J. Toxicol. Environ. Hlth. *2*, 169 (1976)
119. Seguchi, K., Asaka, S.: Intake and excretion of diazinon in freshwater fishes. Bull. Environm. Contam. Toxicol. *27*, 244 (1981)
120. Snegaroff, J.: 7-Ethoxycoumarin deethylation in rainbow trout (*Salmo gairdneri*). Bull. Environm. Contam. Toxicol. *29*, 492 (1982)
121. Gaughan, L.C., et al.: Distribution and metabolism of *trans-* and *cis-*permethrin in lactating Jersey cows. J. Agric. Food Chem. *26* (3), 613 (1978)
122. Sperber, I.: The formation of sulphuric esters of phenols in the chickens. Ann. R. Agric. Coll. Sweden *16*, 446 (1949)
123. Stanton, R.H., Khan, M.A.Q.: Hepatic mixed-function oxidase in freswater fishes: metabolism of and interaction with cyclodiene insecticides and their photoisomers. Amer. Zool. *13* (4), 1308 (1973)
124. Suflita, J.M., et al.: Dehalogenation: A novel pathway for the anaerobic biodegradation of halogenated compounds. Science *218*, 1115 (1982)
125. Suzuki, J., et al.: Formation of mutagens by photochemical reaction of biphenyl in nitrate aqueous solution. Chemosphere *11* (4), 437 (1982)
126. Suzuki, T., Takeda, M.: Microbial metabolism of N-methylcarbamate insecticide. I. Metabolism of o-sec-butylphenyl N-methylcarbamate by *Aspergillus niger* van Tieghem. Chem. Pharm. Bull. *24* (9), 1967 (1976)
127. Suzuki, T., Takeda, M.: Microbial metabolism of N-methylcarbamate insecticide. II. Synthesis of metabolites of o-sec-butylphenyl N-methylcarbamate. Chem. Pharm. Bull. *24* (9), 1976 (1976)

128. Suzuki, T., Takeda, M.: Microbial metabolism of N-methylcarbamate insecticide. III. Time course in metabolism of *o-sec*-butylphenyl N-methylcarbamate by *Aspergillus niger* and species differences among soil fungi. Chem. Pharm. Bull. *24* (9), 1983 (1976)

129. Suzuki, T., Takeda, M.: Microbial metabolism of N-methylcarbamate insecticide. IV. New metabolites of *o-sec*-butylphenyl N-methylcarbamate by *Cladosporium* cladosporioides. Chem. Pharm. Bull. *24* (9), 1988 (1976)

130. Takimoto, Y., Miyamoto, J.: Studies on accumulation and metabolism of sumithion in fish. J. Pestic. Sci. *1*, 261 (1976)

131. Tulp, M.Th.M., Hutzinger, O.: Rat metabolism of polychlorinated dibenzo-p-dioxins. Chemosphere *7* (9), 761 (1978)

132. Uesugi, Y., Sisler, H.D., Metabolism of a phosphoramidate by *Pyricularia oryzae* in relation to tolerance and synergism by a phosphorothiolate and isoprothiolane. Pestic. Biochem. Physiol. *9*, 247 (1978)

133. Walter-Echols, G., Lichtenstein, E.P.: Microbial reduction of phorate sulfoxide to phorate in a soil-lake mud-water microcosm. J. Econ. Entomol. *70*, 505 (1977)

134. Walter-Echols, G., Lichtenstein, E.P.: Movement and metabolism of [^{14}C]phorate in flooded soil system. J. Agric. Food Chem. *26* (3), 599 (1978)

135. Walter-Echols, G., Lichtenstein, E.P.: Effects of lake bottom mud on the movement and metabolism of ^{14}C-phorate in a flooded soil-plant system. J. Environ. Hlth. *13* B (3), 149 (1978)

136. White, I.N.H., Mattocks, A.R.M., Butler, W.H.: The conversion of the pyrrolizidine alkaloid retorsine to pyrrolic derivatives *in vivo* and *in vitro* and its acute toxicity to various animal species. Chem.-Biol. Interactions *6*, 207 (1973)

137. Willis, D.E., Addison, R.F.: Hydroxylation of biphenyl *in vitro* by tissue preparations of some marine organisms. Comp. Gen. Pharmac. *5*, 77 (1974)

138. Wilson, T.G., Gilbert, L.I.: Metabolism of juvenile hormone I in *Drosophila melanogaster*. Comp. Biochem. Physiol. *60*A, 85 (1978)

139. Wit, J.G.: Drug enzymes in kestrel liver. Comp. Biochem. Physiol. *30*, 185 (1969)

140. Yu, S.J., Terriere, L.C.: Metabolism og [^{14}C]hydroprene (ethyl 3,7,11-trimethyl-2,4-dodecadienoate) by microsomal oxidases and esterases from three species of diptera. J. Agric. Food. Chem. *25* (5), 1076 (1977)

141. Yu, S.J., Terriere, L.C.: Ecdysone metabolism by soluble enzymes from three species of diptera and its inhibition by the insect growth regulator TH-6040. Pestic. Biochem. Physiol. *7*, 48 (1977)

142. Yu, S.J., Terriere, L.C.: Metabolism of juvenile hormone I by microsomal oxidase, esterase, and epoxide hydrase of *Musca domestica* and some comparisons with *Phormia regina* and *Sacrophaga bullata*. Pestic. Biochem. Physiol. *9*, 237 (1978)

143. Yumita, T., Shoji, A., Yamamoto, I.: Metabolism of Mepronil (Basitac) in rice plants. J. Pestic. Sci. *6*, 347 (1981)

144. Van Duuren, B.L.: Prediction of carcinogenicity based on structure, chemical reactivity and possible metabolic pathways. J. Environ. Pathol. Toxicol. *3*, 11 (1980)

145. Smith, G.N., Thiegs, B.J., Ludwig, P.D.: The isolation and identification of the amino-nitro-o-toluamide formed by the biological reduction of 3,5-dinitro-o-toluamide. J. Agric. Food Chem. *11*, 257 (1963)

146. Cerniglia, C.E., et al.: Metabolism of benzidine and benzidine-congener based dyes by human, monkey and rat intestinal bacteria. Biochem. Biophys. Res. Commun. *107* (4), 1224 (1982)

147. Gaughan, L.C., Unai, T., Casida, J.E.: Comparative permethrin metabolism: Rat and cow, bean and cotton. Abstr. 172nd ACS Meeting, San Francisco, CA, PEST 36 (1976)

148. Gaughan, L.C., Unai, F., Casida, J.E.: Permethrin metabolism in rats. J. Agric. Food Chem. *25* (1), 9 (1977)

149. Gaughan, L.C., Robinson, R.A., Casida, J.E.: Distribution and metabolic fate of *trans*- and *cis*-permethrin in laying hens. J. Agric. Food Chem. *26* (6), 1374 (1978)

150. Lynn, R.K., et al.: Disposition of the flame retardant, tris(1,3-dichloro-2-propyl) phosphate, in the rat. Drug Metab. Dispos. *9* (5), 434 (1981)

151. Lynn, R.K., et al.: Metabolism, distribution and excretion of the flame retardant tris(2,3-dibromopropyl)phosphate (Tris-BP) in the rat: identification of mutagenic and nephrotoxic metabolites. Toxicol. Appl. Pharmacol. *63*, 105 (1982)

152. Thomas, J.M., Nakaue, H.S., Reid, B.L.: Effect of acetylsalicylic acid and some chemical analogs on chick growth and selected biochemical parameters. Poult. Sci. *46*, 1216 (1976)
153. Fingl, E.: Laxatives and cathartics. *In* Pharmacological Basis of Therapeutics, ed. L.S. Goodman and A. Gilman, Maxmillan Publishing Co. (1980)
154. Van Duuren, B.L., Loewengart, L.: Reaction of DNA with glycidaldehyde. J. Biol. Chem. *252*, 5370 (1977)
155. Ohta, T., et al.: Photochemical nitrosation of dimethylamine in aqueous solution containing nitrite. Chemosphere *11* (8), 797 (1982)

Subject Index

Environmental Toxin Series

Editors:
S. Safe, O. Hutzinger

Springer-Verlag
Berlin Heidelberg New York
London Paris Tokyo

Volume 1

Polychlorinated Biphenyls (PCBs): Mammalian and Environmental Toxicology

With contributions by numerous experts

1987. 33 figures, 35 tables. X, 125 pages.
ISBN 3-540-15550-3

Contents: *S. Safe, L. Safe, M. Mullin:* Polychlorinated Bi-
phenyls: Environmental Occurrence and Analysis. –
L. G. Hansen: Environmental Toxicology of Polychlorinated
Biphenyls. – *A. Parkinson, S. Safe:* Mammalian Biologic and
Toxic Effects of PCBs. – *M. A. Hayes:* Carcinogenic and
Mutagenic Effects of PCBs. – *I. G. Sipes, R. G. Schnellmann:*
Biotransformation of PCBs: Metabolic Pathways and
Mechanisms. – *R. J. Lutz, R. L. Dedrick:* Physiologic Phar-
macokinetic Modeling of Polychlorinated Biphenyls. –
S. Safe: PCBs and Human Health. – Subject Index.

Volume 2

Cadmium

3rd IUPAC Cadmium Workshop, Jülich, Federal Republic
of Germany, August 1985

With contributions by M. Stoeppler and M. Piscator

1988. 57 figures, 79 tables. ISBN 3-540-15551-1

Contents: Toxicity, Carcinogenicity, Animal Experiments.
– Epidemiology. – Cadmium in the Environment. –
Methodology and Quality Assessment.

The concern about environmental toxins is ever increasing,
as is the need for sound scientific information. The Environmental Toxin Series is dedicated to the publication of
comprehensive reviews and monographs on compounds or
classes of chemicals which are of importance in environmental toxicology. The series is designed to serve as a
background of information for scientific investigation as
well as risk analysis and political decision making. The
main aim of the series is to describe in as complete a way
as possible all potentially hazardous chemicals from the
point of view of chemistry, ecology, toxicology, risk analysis
and regulatory implications. From time to time conference
proceedings on important and urgent topics will be
included in the series. We thank the members of the edito-
rial board for their enthusiastic support.

Springer

W. Fresenius, K.-E. Quentin, W. Schneider
(Eds.)

Water Analysis

**A Practical Guide to Physico-Chemical,
Chemical and Microbiological Water
Examination and Quality Assurance**

Compiled by W. Schneider

Published by Deutsche Gesellschaft für Technische
Zusammenarbeit (GTZ), Eschborn (FRG)

With contributions by F. J. Bibo, H. Birke, H. Böhm,
W. Czysz, H. Gorbauch, H. J. Hoffmann, H. H. Rump
with Additional Contributions by Other Experts

Translated from the German by A. Gledhill, R. Holland,
T. J. Oliver

1988. 178 figures, 55 tables. XXV, 804 pages.
ISBN 3-540-17723-X

Contents: Introduction, Sampling, Local Testing, etc. –
Theoretical Introduction to Selected Methods of Water
Analysis (Classic and Instrumental Methods. – Inorganic
Parameters. – Organic Parameters. – Biological Analysis.
– Evaluation of Analysis Data. – Subject Index.

The book is written for the practitioner in water analysis
and quality assurance of drinking water. In addition to
detailed instructions for sampling and immediate analysis, the book provides a concise presentation of the theoretical background and of data evaluation. The presented
analytical methods can be applied successfully with
simple equipment as well as in modern advanced laboratories.
The book is a bench-top laboratory manual and can be
used for instruction in laboratory staff training programs.
It treats the analysis of organic and inorganic compounds
and also deals with microbiological problems associated
Springer-Verlag   with the guidelines for waste water, surface and ground
Berlin Heidelberg New York   water, and drinking water quality.
London Paris Tokyo

Springer